S0-AKJ-165

ACS SYMPOSIUM SERIES **478**

Partnerships in Chemical Research and Education

James E. McEvoy, EDITOR
Consultant, Industrial–Academic Relations

Developed from a symposium sponsored
by the Division of Industrial and Engineering Chemistry, Inc.
at the 200th National Meeting
of the American Chemical Society,
Washington, DC,
August 26–31, 1990

American Chemical Society, Washington, DC 1992

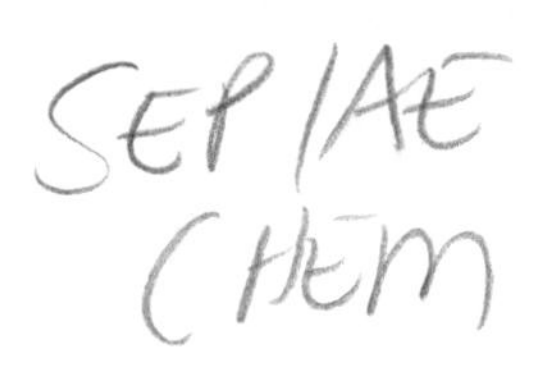

Library of Congress Cataloging-in-Publication Data

Partnerships in chemical research and education/ James E. McEvoy, editor

p. cm.—(ACS Symposium Series, 0097–6156: 478).

"Developed from a symposium sponsored by the Division of Industrial and Engineering Chemistry, Inc. at the 200th National Meeting of the American Chemical Society, Washington, D.C., August 26–31, 1990."

Includes bibliographical references and indexes.

ISBN 0–8412–2173–1

1. Chemistry—Research—Congresses. 2. Chemical engineering—Research—Congresses. 3. Industry and education—Congresses. I. McEvoy, James E., 1920– . II. American Chemical Society. Division of Industrial and Engineering Chemistry. III. American Chemical Society. Meeting (200th: 1990: Washington, D.C.). IV. Series.

QD40.P293 1991
540′.72—dc20 91–32420
CIP

The paper used in this publication meets the minimum requirements of American National Standard for Information Sciences—Permanence of Paper for Printed Library Materials, ANSI Z39.48–1984. ∞

PRINTED IN THE UNITED STATES OF AMERICA

ACS Symposium Series

M. Joan Comstock, *Series Editor*

Foreword

THE ACS SYMPOSIUM SERIES was founded in 1974 to provide a medium for publishing symposia quickly in book form. The format of the Series parallels that of the continuing ADVANCES IN CHEMISTRY SERIES except that, in order to save time, the papers are not typeset, but are reproduced as they are submitted by the authors in camera-ready form. Papers are reviewed under the supervision of the editors with the assistance of the Advisory Board and are selected to maintain the integrity of the symposia. Both reviews and reports of research are acceptable, because symposia may embrace both types of presentation. However, verbatim reproductions of previously published papers are not accepted.

Contents

Preface

THIS BOOK IS ABOUT PARTNERSHIPS in chemistry and chemical engineering. Although such partnerships are by no means novel, there has been an evolution toward multifaceted arrangements that include industry, academia, federal and state governments, and precollege educators. Cooperative partnerships between universities and industrial organizations have been developing at an accelerating rate for the past 15 to 20 years; however, it has been only during the last 5 years that we have seen spectacular growth in the involvement of academia and industry in precollege education arrangements. The reasons for this growth include many factors that reflect concern, such as the

- declining enrollments of science and engineering majors at the university level.
- lack of interest and excitement in science among precollege students.
- inadequate training of science teachers in elementary and secondary schools.
- negative perceptions of chemistry and chemicals by the population in general.

This book provides descriptions of many of the imaginative programs that have been developed in recent years both to provide support for basic research in universities and to stimulate interest in chemistry at the precollege level. The chapters were developed from papers presented at two symposia given at national meetings of the American Chemical Society: *Synergistic Programs in the Chemical Sciences and Engineering* organized by me and presented in August 1990, and *Industrial Initiatives in Precollege Science Education* organized by C. Gordon McCarty and Kenneth O. MacFadden and presented in April 1990.

We are hopeful that the programs described herein can serve as models for forming useful partnerships among groups similar to those represented in this book. *Partnerships in Chemical Research and Education* should be helpful to all those persons and organizations concerned with chemistry in the United States because this is where it all begins—in the primary and secondary grades, the colleges, the universities, and the academic and industrial research laboratories.

My thanks to all the contributors for their cooperation in meeting publication deadlines and to those persons who participated in the corresponding symposia, as well. I also acknowledge the help of Cheryl Shanks, ACS Books, and her invaluable advice during the editorial process; Ken Chapman, ACS Division of Education, for his encouragement; and Nancy Flinn, Manager of the Corporation Associates, for her assistance in providing access to the papers from the April 1990 symposium. The generous financial support of E. I. du Pont de Nemours and Company and Hercules, Inc., is also acknowledged. These contributions have helped to decrease the selling price of this book in order to make it more available to the education community.

JAMES E. MCEVOY
Industrial–Academic Relations
Bethlehem, PA 18015

August 6, 1991

Introduction
by Paul G. Gassman

The Scientific Pipeline in Chemistry: Working Together To Fill the Needs of Academia and Industry

Understanding the interdependency of industry and academia, our mutual needs, the problems we face in the future, and the goals we will need to accomplish must be our primary concerns if we are to have a future working together.

Mutual Need

Let us consider what is happening in the United States as relates to chemistry. There are presently about 4.6 million scientists and engineers. Of those, 3.2 million (69%) are employed by industry, and 600,000 of those are engaged in research and development. Research and development is where the future lies, and that's where a great deal of cooperation, interchange, and interaction are needed if we are to have a viable future.

A closer look at those 600,000 scientists and engineers shows that 360,000, or slightly more than half, are scientists, and 240,000 are engineers. (These statistics were compiled by the Industrial Research Institute, which also examined the subdisciplines.) Surprisingly, 50 percent of all scientists and engineers involved in research and development in the industrial sector are either chemical engineers or chemists. There is no doubt but that this particular group contributes greatly to the well-being of the chemical industry and to the general well-being of industry in this country. When one looks at these numbers, it is not surprising that in 1989, the chemical industry had a $15.7 billion positive balance of trade and, in 1990, a $16.2 billion surplus was obtained. Industry is based heavily on research and heavily on developing new and innovative technology. Future problems will most likely come, in part, from the lack of a sufficient supply of chemists.

In order to maintain a strong technological base—and without a strong technological base the industry will not be able to maintain its superiority in the field of chemistry—chemical technicians, B.S. chemists, M.S. chemists, Ph.D. chemists, and B.S., M.S., and Ph.D. chemical engineers are needed. Chiefly, as pertains to the development of technology and innovative processes, we will need a constant supply of Ph.D. chemists, a "pipeline." In terms of producing Ph.D. chemists, there are many problems. The pipeline problem has not yet become apparent, but it's easy to see what is coming. There is no light at the end of the pipeline, and what can be seen from here is rather dismal.

The "Pipeline"

There are 23,000 high schools in the United States. Of those, 7,000 offer no physics courses, 4,200 offer no chemistry courses, and 1,900 offer no biology courses. If we look at other things that are happening in terms of high school laboratory experiences, things are going downhill at a very fast rate. Most students get turned on to chemistry by *doing* things in the laboratory with their own hands, not by listening to lectures either in high school, or in college, or even in graduate school, because some of those are awfully dull.

According to the preamble to Sen. Edward Kennedy's 1990 Omnibus Education Bill, 53% of our high schools in 1977 offered a laboratory to students in at least one science. By 1989, just 12 years later, the percentage of high schools providing students with the opportunity to take a laboratory course had dropped from 53% to only 39%. That is a frightening statistic.

We have all seen the statistics on how bright—or how unbright, depending on how one chooses to look at it—our graduating high school seniors are. According to a 1988 survey of science students achievements (graduating high school seniors), entitled "Science Achievement in 17 Countries," the United States ranked 13th of 13 countries surveyed in biology, 11th of 13 in chemistry, and 9th of 13 in physics. (The countries surveyed are Australia, Canada, England, Finland, Hong Kong, Hungary, Italy, Japan, Norway, Poland, Singapore, Sweden, and the United States.) This situation is bound to get worse. Between 1990 and the turn of the century, half of the high school science teachers in the United States will retire. That is another frightening statistic. Very few students are being trained to be high school science teachers. The supply of science educators coming down the pipeline is diminishing rapidly, without being replenished.

A look at the supply of new chemists (*see* page xi) for the years 1974–1985 (chemists with bachelor degrees are shown in the upper bars of the graph), shows that 1978 was the peak, with a definite downward

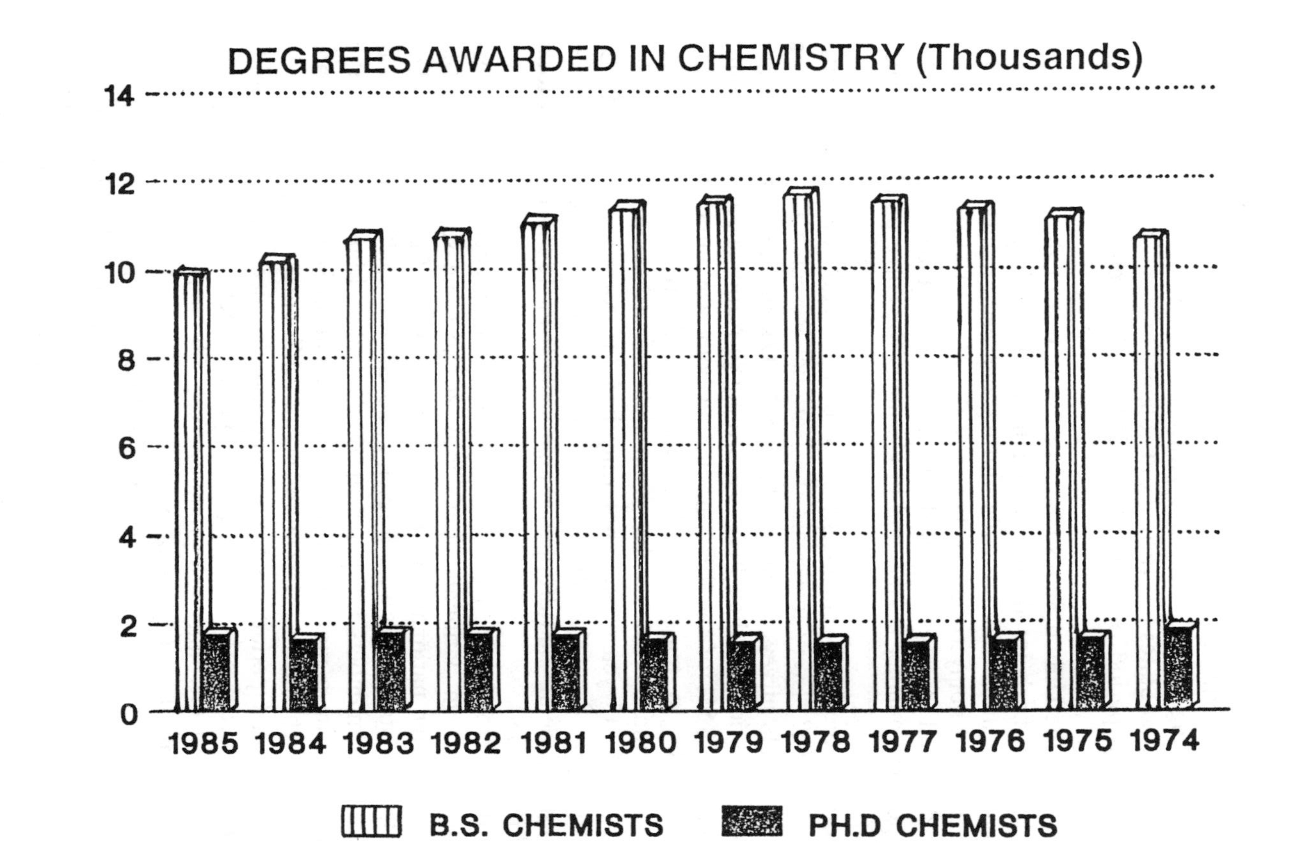
DEGREES AWARDED IN CHEMISTRY (Thousands)
14
12
10
8
6
4
2
0
1985
1984
1983
1982
1981
1980
1979
1978
1977
1976
1975
1974
B.S. CHEMISTS
PH.D CHEMISTS
Source: Department of Education's National Center for Educational Statistics

momentum since then. Across the bottom are bars that represent Ph.D. production; that remains fairly constant. The years 1984–88 show a downward progression in B.S. graduates, from 9,800 to 8,200. For the past 10 years, our production of B.S. chemists has dropped roughly 3% per year. For further comparison, consider these statistics: In the past 36 years (according to the National Research Council Doctoral Records File), the United States has turned out roughly one million lawyers. During that same period of time, about 55,000 Ph.D. chemists were graduated.

How does the United States compare to the rest of the world in terms of turning out bachelor degrees in sciences? According to the Kennedy Omnibus Education Bill, students earning degrees in physics, chemistry, math, and astronomy (what we know as the physical sciences) represent only 5% of all degrees awarded in the United States. In Japan the corresponding number is 20% and in Germany the corresponding number is 37%.

Where do the Ph.D. chemists intend to work, and how are their attitudes changing? Every five years the National Research Council does a fairly thorough survey of the graduating Ph.D.s in chemistry and what they intend to do with their lives. According to National Research Council statistics, 18.5% of the 1970 graduating body was interested in going into academia, to teach future scientists, to do research, and to get involved in fundamental science. That 18.5% translates to 411 individuals. By 1985, 15 years later, not only had the total number of Ph.D. chemists dropped considerably, but the percentage that said they wanted to go into academia had dropped dramatically from 18.5% to 8%, a total of 147 individuals.

Another factor impinging on the supply of Ph.D. chemists is the percentage of foreign nationals. Let's look at the same survey to see the impact of foreign nationals. In 1970, the percentage of graduating Ph.D.s in the United States who were foreign nationals was 15.2%; in 1985 it was 22.7%. Those numbers are going up dramatically. Many schools in the United States have more than 50% foreign nationals in their graduate programs in chemistry. The number of U.S.-born Ph.D.s decreased steadily over that period—and it is continuing to decrease. In the mid-1990s the United States will graduate approximately as many U.S.-born Ph.D.s as West Germany will graduate West German-born Ph.D.s (prior to reunification of East Germany with West Germany).

It is also important to look at the pipeline in terms of its length. In 1967, the total time given to earning the doctorate (the time elapsed between receiving a bachelor's degree and earning a Ph.D. in chemistry) was 6.36 years; in 1986 it was 7.20 years. An additional factor to consider is the "mean registered time"; in other words, the average time the student is actually enrolled in graduate school. That has gone from 5.01

years in 1967 to 5.83 years in 1986. (This data comes from "On Time to the Doctorate," published in 1990 by the National Academy of Science Press.)

One of the main problems with the pipeline is that it cannot be shortened quickly, because it affects not only bachelor students and Ph.D. graduate students but junior and senior high school students. In effect the pipeline is about 18 years long. We could shorten that by perhaps a year or two with extensive finetuning, but we cannot shorten it significantly. The demand—currently about 200 faculty positions open each year because of retirements alone—is pretty well established.

A 1987 survey by Bergman and Heathcock (Bergman–Heathcock Report to the National Science Foundation) reviewed all Ph.D.-granting institutions in the United States (response rate: 83%), and found that 316 academic vacancies could be identified. When the results were corrected to 100%, the figure was 381 openings. One of the most interesting things Bergman and Heathcock found was that there are numerous chronic vacancies—vacancies that have been open an average length of 2.5 years. In the schools that responded there were 133 such vacancies. Extrapolating that to include all Ph.D.-granting institutions gives a figure of 160 chronic vacancies. It seems that we are producing less than half the number of people needed each year. Each August, there are about 400 openings in academia. The chemistry faculty for B.S.-, M.S.-, and Ph.D-granting institutions in the United States numbers about 12,000, so 400 openings each August is not surprising. In fact, it's pretty reasonable. With an average career length of 30 years (which is probably a bit on the long side), 400 new positions would open each year. In 1985, 1,836 Ph.D. chemists were graduated. Only 147 of those expressed an interest in teaching at any level.

While there are at present about 400 openings, the number of openings is increasing. The problem is that the numbers of individuals retiring are increasing very rapidly. In about four years the slope on retirements increases rapidly. In four years about 300 new openings each year will be generated, in about nine, 500 new openings per year; and in about 12 years we peak at 600 new openings per year due to retirements alone. That, combined with the fact that only 100 to 200 individuals in any given year want academic jobs, means that the backlog of openings is going to increase very rapidly. It seems fairly safe to project that near the turn of the century, there will be approximately 2,000 open positions on chemistry faculties throughout the country.

The beginnings of this are already visible, and its impact in terms of salaries and set-up packages for new faculty is already being felt. At some major universities, salaries based on a 12-month period for new assistant professors are higher by a significant amount than starting salaries in industries. Over a 10-year period, set-up packages have gone up roughly by an order of magnitude. From 1960 to 1980, the increase was very small, but that was supply and demand. However, during the 1980s, set-up packages increased dramatically (*see* page xv). One set-up package of $780,000 was provided for a new assistant professor last year by an east coast school. One year later, the top set-up package for a new assistant professor was reported to be in excess of $1 million. The compression of salaries in industry is occurring at a very rapid rate. This is because of small increases for longtime employees and higher starting salaries for new employees. Last year, two companies on the east coast gave selected new Ph.D. chemists $10,000 signing bonuses. One east coast company has recently been providing $20,000 gifts for home purchases, as long as the individual agrees to remain with the company for three years. It's a loan for the first three years, after that, it's a gift. These companies are competing for the best.

It is hard to imagine how long the system can afford to continue in this vein. This would seem to be the perfect example of supply and demand. It's Economics 101 in its simplest form: Supply dwindles, demand increases, prices go up. There must be a point at which the price stops going up. According to the formula, the supply should begin to increase as more people are attracted to the market's higher prices. The problem is that the pipeline is so long that the supply cannot increase significantly in less than a decade.

Let's look at the demographics of the chemistry faculty (doctoral) in the United States. In 1975, the bulk of faculty was between 30 and 35 years old. Ten years later, the largest group was aged 40 to 45. University faculty are no longer staying on until age 70 but are retiring earlier in ever increasing numbers, without being replaced proportionately by incoming faculty. In part, this is because many universities have put together improved retirement packages. When faculty members get to the point where their retirement pay will equal their salary, it takes very little to prod them into leaving—a simple argument with the chairman or the dean can be the impetus that sends them out the door. In another 12 years, retirements will hit their peak, generating as many as 600 openings per year.

A tremendous number of retirements are coming up, and the supply of replacements is going to be quite limited. The main problem is that the number of individuals obtaining Ph.D. degrees in the United States per 1,000 population has decreased dramatically. According to the National Academy of Science, in 1971, 1.6 of every 1,000 U.S. citizens was earning a Ph.D. in physics, chemistry, or mathematics. By 1985, that number had dropped to below 0.6 per 1,000. The same sort of phenomenon is occurring in engineering. In 1971, 0.9 of every 1,000 U.S. citizens was earning a Ph.D. in engineering; by 1985 the ratio had dropped to 0.3 per 1,000.

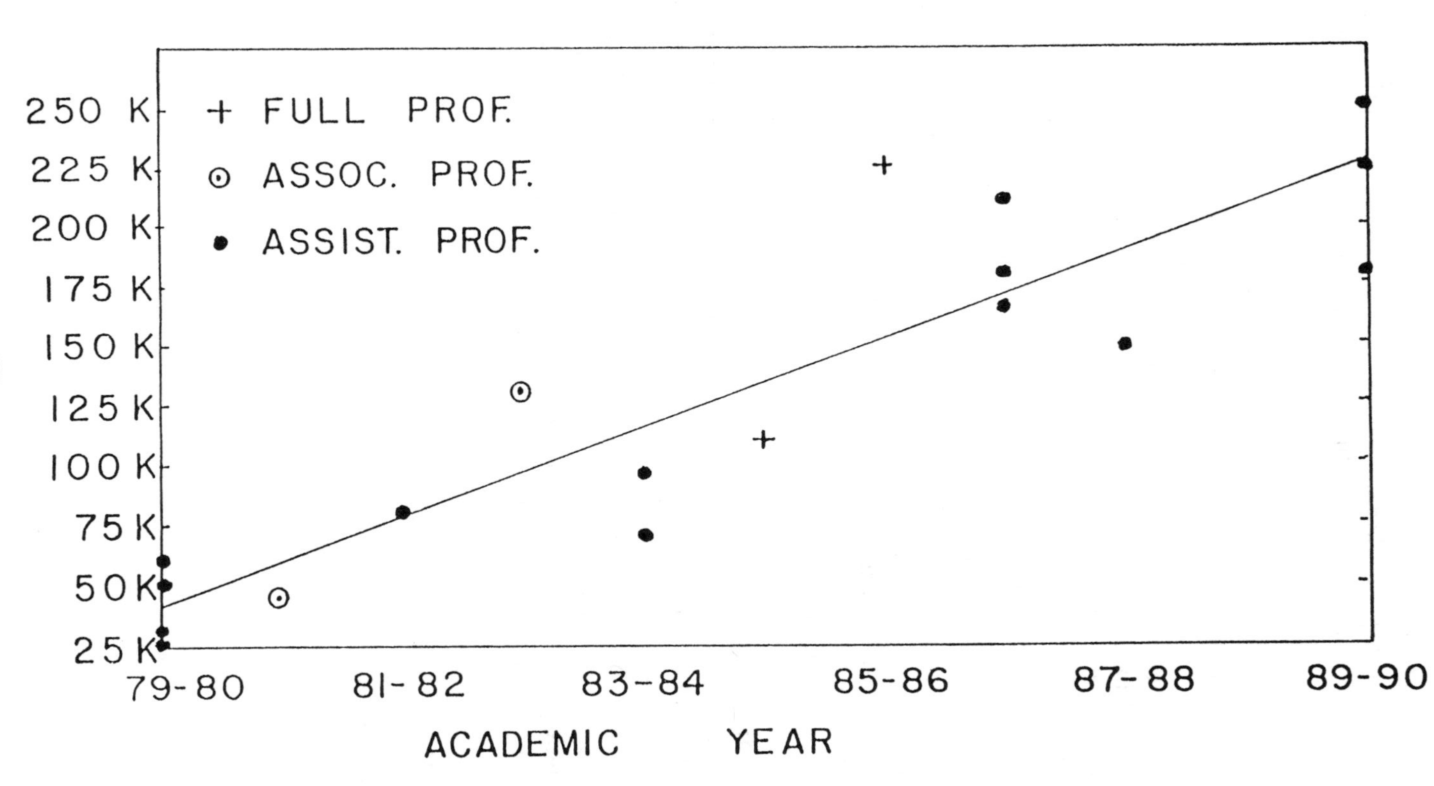
SET-UP FUNDS FOR NEW FACULTY — UNIV. of MINNESOTA
+ FULL PROF.
⊙ ASSOC. PROF.
• ASSIST. PROF.
250 K
225 K
200 K
175 K
150 K
125 K
100 K
75 K
50 K
25 K
79-80
81-82
83-84
85-86
87-88
89-90
ACADEMIC YEAR

These numbers are still declining a bit and, although the drops are not as dramatic as they have been, it is clear that we have a very serious problem in terms of attracting the youth of America to science and engineering.

The Future

Without strong cooperation between academia, industry, and government, we will not have the scientists needed now and in the future to remain competitive on a global basis. We will not be able to maintain the technological base we need to maintain in order to be competitive with Germany, which has a fantastic chemistry industry, with Japan, which has a strong chemical base, or with developing nations. Spain is the fastest-growing country in Europe in terms of papers being abstracted and patents being abstracted by Chemical Abstracts. Traditionally, two-thirds of the foreign nationals who come to the United States for graduate-level education remain in the United States. That will change as the chemical industry develops throughout the world. Individuals who might formerly have studied and stayed in the United States may find it more attractive to study and stay elsewhere or to return to their native countries for even better opportunities.

The bottom line for the future is that we must form coalitions among industry, academia, and government that will allow us to (a) attract more and better students into science, (b) provide better training through grade school, high school, college, and graduate school, and (c) develop an industrial sector that prides itself on maintaining a continuing, cutting-edge education mode for young scientists. We must change the pipeline, or we will not have a supply of scientists who will be involved in those programs. We must cooperate not only in technology transfer but in generating the people who can and will do that technology.

PAUL G. GASSMAN
Department of Chemistry
University of Minnesota
Minneapolis, MN 55455

Chapter 1

Synergy in Chemical Research and Education

James E. McEvoy

Consultant, Industrial–Academic Relations, 1875 Quarter Mile Road, Bethlehem, PA 18015

During the last ten years there has been a significant increase in positive interactive relationships in the chemical sciences and engineering between industry, academia and federal and state government. This chapter describes the evolution of these programs, obstacles to further development and the hope for future progress.

This book has been written by persons actively involved in development of a multitude of interactive programs that are taking place in the public and private sectors of our national chemical enterprise. Hopefully these examples will provide inspiration to others to develop their own approaches to making science and particularly chemistry more relevant to our nation and provide a basis for a better future for those who will come after us.

In order to understand the significance of this collection of papers it is necessary to look at the immediate period following World War 2 and also to examine the status of these interactions during the prewar period. Books, or for that matter publications of any kind, containing descriptions of interactive programs would have been virtually impossible to find. Prewar industrial support for academic research was fairly common but practised primarily by the larger chemical and petroleum companies. Federal support was minimal during the prewar period but as the federal bureaucracy expanded after the war so did the support for basic research in the sciences and engineering. In chemistry the principal federal agency responsible for this growth was the National Science Foundation . This was indeed fortuitous as industrial organizations, although expanding their research and development budgets overall, were cutting back on their internal investment in basic research in favor of the shorter term oriented applied R&D. Focus in the private sector was on NPAT, (net profit after taxes) and annual budgetary planning for the profit centers of most major corporations started here and worked

0097–6156/92/0478–0001$06.00/0

backwards to determine the inputs necessary to achieve the specified financial goals. Long range research did not have a high priority in this environment and gave way to the near term (3-5 year) alternatives. Currently "true" basic research in the chemical sciences is almost exclusively being done by the universities.

In the chapter " The Council for Chemical Research : Developing the Trust Factor ", Ivan Legg , Dean of the Colleges of Sciences and Mathematics at Auburn University mentions the beginnings of CCR in 1980. The initial impetus for this organization came from "Mac" Pruitt at that time , (1979) , the Vice President for Research for the Dow Chemical Company. "Mac" was very concerned about the level of basic chemical research that was being done in the United States and believed that the decline in this activity was a threat to our national pre-eminence in this science. Accordingly he proposed that if U.S. industry was reluctant to invest in basic research, it should at least be supporting research in the universities. This support was not to be restricted to financial aid but would require a high degree of interaction between industrial and university chemists and chemical engineers.

It is interesting to note that as CCR grew, financial support of university research from industry grew rapidly through direct interaction of industrial persons with their academic counterparts rather than through industrial contributions to a general fund with subsequent distribution from the parent organization to the universities. During the organizing phase of CCR a debate developed between those favoring establishment of a fund to support research through substantial contributions of industrial members to a central fund to be managed by CCR according to a predetermined formula and those favoring the direct interaction of the academic and industrial members on a "one-on-one" basis for establishment of mutual research interests . Eventually the Chemical Sciences and Engineering Fund (CSEF) was established to receive contributions on a voluntary basis to be distributed according to the number of PhD's being graduated at member universities. This fund never grew to the size that the founding group originally hoped for and in fact reached a steady state and has been in decline for the past few years. During this same period direct industrial support to university chemistry and chemical engineering grew substantially.. (> 15%/year in the period 1984-1988). There remains little doubt that industrial financial support is not purely altruistic. There is nothing wrong with this as long as the academicians maintain their professional integrity . There is no evidence to indicate otherwise.

We have also seen a rapid increase in industrial and academic interaction in precollege education during the last decade. The depth and breadth of this interaction is shown in the response to a call for for papers by the ACS Corporation Associates for an ACS symposium which took place in Boston, Mass. in 1989 titled " Industrial Initiatives in Precollege Sc ience Education". 18 major companies presented papers describing their direct involvement in precollege programs. which was directed at the entire spectrum of elementary, schools, middle schools , high schools and teacher training . Industrial scientists visited the schools and educators participated in industrially organized programs in the workplace. This interest was triggered in part by the realization that the The United States was not the world leader in math and science education. Possibly of more immediate concern was the fact that the number of students entering universities that were deciding on math and science careers were declining. There were many factors involved in this dismal state of affairs. Principal among these was the perception that math and science were difficult subjects and not very interesting. The inadequate training of primary and secondary school teachers was another contributing factor. It was also recognized that the interest of minorities in math and science careers was declining at the same time that projections on the rapid growth of ethnic minorities in the U.S. were being made.Where were our future scientists --- industrial and academic--- going to come from?

The significant contribution of the federal and state governments to these collaborative programs should be noted . The federal government in particular has recognized the importance of basic research and has continued to maintain financial support for this area even during years of budgetary constraint. This is particularly evident in the chemistry portion of the NSF and DOE budgets which in spite of significant reductions in applied research and demonstration projects , has increased spending for basic research during the last decade. On the other hand State support of research budgets has focused more on the applied shorter term research and development programs. Excellent examples of State/ Industrial/ Academic cooperation in research are the Ben Franklin Partnerships developed over the last decade in Pennsylvania.. Here we see the state of Pennsylvania along with Pennsylvania industry and state and private universities cooperating in research, development and financing to assist fledgling science oriented businesses which eventually could contribute to growth of the state economy.

There are of course obstacles to be overcome. Common to virtually all of these programs is the problem of obtaining financial support. Industrial and government support is the main source of funding for most of these interactive programs. Tighter reins on discretionary industrial contributions have been the order of the day during the recent recessionary period. There are of course a few cynics who see collaboration or interaction of any kind between industry , academia, or the pre-college school system as eminently evil or self serving. There is no denying the fact that "enlightened self interest", is a factor in industrial participation in these programs. It is important for industry to insure a future source of technically trained people to maintain their own competitiveness in a highly competitive society both domestically and internationally. Good community relations are also important . Employees and their families are part of the community and participate in local government and civic functions. Company support for these activities is beneficial both to the community and company, or for that matter the college, university , or institution .

If we are not already directly involved in these programs, perhaps we can at least be supportive of our colleagues in speaking out in those areas of our professional expertise that will heighten our image as chemists or what chemistry produces. The future for our profession depends on what we do now--- working together to insure that our nation will remain pre-eminent in the chemical sciences and engineering.

RECEIVED July 15, 1991

Chapter 2

The Council for Chemical Research

Developing the Trust Factor

J. Ivan Legg

College of Sciences and Mathematics, Auburn University, Auburn, AL 36849–5319

In the late seventies it was evident that we were beginning to lose our technological lead in the world economy. In an attempt to address this issue, Mac Pruitt, Research VP at Dow, convened a meeting of academic and industrial leaders from the chemical sciences and engineering community. It was clear that the vital link between the two parties was in critical need of improvement, and the Council for Chemical Research was formed. CCR has had a positive impact over the past decade. CCR's success is based on the trust that has evolved between the academic and industrial members who represent the major part of the research leadership in chemical sciences and engineering. CCR is truly unique among scientific and engineering organizations in the U.S.

Consider the following scene, a scene that will bring warm memories to some of you. It is a late Saturday afternoon. You walk into your research laboratory to find your group hard at work. A graduate student is listening intently to one of your postdocs explain the fine points of preparing a column for chromatography. Another student is pouring over an nmr of a compound you need to complete a manuscript. Two students are arguing at the blackboard over the interpretation of analytical data. In the background strains of Beethoven's 6th symphony, well, maybe Dire Straights' popular hit, "Money for Nothing . . .", come from an old tape deck installed many years ago by a former student who is now a successful professor at a respectable university.

You smile and return to your office to put the finishing touches on a manuscript that promises to have a major impact in your field. You have much to celebrate today because your program director called yesterday to inform you that your grant had been renewed.

0097–6156/92/0478–0005$06.00/0

A few weeks later a colleague from industry, after carefully reading a preprint of your seminal paper, comes for a visit and decides that a long term investment in your project is a wise decision. The company invests, and five years later a patent is filed that promises riches for all.

The curtain drops, with the lyrics of the Dire Straights' hit "Money for Nothing . . ." strumming away in the background.

Too good to be true? Wishful thinking? Most of us would answer yes. All is not well in the world of science and technology; and in some way, the relationship between industry, academia, and government comes up again and again as a major source of the problem.

The Council for Chemical Research, CCR, is addressing the complex problems that places this happy story in the category of myth. The Council's makeup positions it uniquely to address the issues that stand in the way of a productive relationship between academia and industry as detailed below. But first some provocative statements and some history that will set the stage for what CCR has been able to accomplish towards improving the industrial/academic interface.

Contrary to popular belief industry and academia are working together to improve the welfare of this country. Contrary to popular belief industry is making a major effort to clean up the environment. Contrary to popular belief man made chemicals such as pesticides are not primary sources of cancer in man. Unexpected statements from an academic? Eleven years ago, yes. Today, no.

Eleven years ago, Mac Pruitt, then Vice President for Research at Dow Chemical, brought together leaders from academia and industry concerned with research and development in the chemical sciences and engineering. Pruitt observed that at that time, "there was a distinct feeling in the scientific community . . . that we were about to lose our technology lead in the world." He believed that one of the main reasons for this was that "industry and universities were no longer cooperating. In fact, they had almost become antagonistic to each other." Pruitt called the meeting in an attempt to determine the extent of the problem and to seek a solution.

He made the following observation of those gathered in September 1979 in Midland, Michigan.

> From the very beginning, you could see that we did have a real problem because everybody, from the university particularly, had a big question: 'What are these guys up to now? Anybody who would invite us here and pay our way, must be up to some scheme.' There was an aura of suspicion or wonderment all through the conference . . . . Most of the people had never met each other.

Up to this time many people had been running around the country, meeting and wringing their hands over the ensuing crisis in science and technology, but nothing much had happened. Pruitt decided that we needed to go beyond the Midland meeting and the Council for Chemical Research was incorporated in late 1980. The three interlocking components of CCR's logo shown in Figure 1 represent academia, government, and industry.

Figure 1. The CCR logo.

CCR has come a long way since then. Although CCR cannot claim with certainty to be a primary player in helping maintain the health of a vast and far reaching industry, it has had an impact, and its potential for further contributions is significant.

CCR is unique in the United States. The organization brings together leaders from academia and industry in an area crucial to the nation's future. Research in the chemical sciences and engineering is central to achievements in the medical sciences, biotechnology, materials, and pollution abatement and is an integral part of one of the few industries in this country with a large, favorable trade balance ($15.7 billion in 1989). The only other major industry left in the United States with a favorable trade balance is the aerospace industry, primarily represented by Boeing. U.S. chemical sales in 1989 were $280 billion. The industry employs over 800,000 people, a number exceeded only in the auto and textile industries.

As noted by Philip Abelson in an editorial in the July 20, 1990, issue of Science:

> Chemistry is a discipline central and essential to the other natural sciences, most of technology, and much of medicine. In the future global competition, a country not tops in chemistry is destined to be second-rate or worse.

CCR's membership consists of chemical engineering and chemistry department heads, deans, and vice presidents from 162 of the nation's leading universities and research directors and vice presidents from 50 major industrial firms. The common denominator in the organization is shared responsibility in leadership of research and education in the chemical sciences and engineering.

The Council for Chemical Research is lead by a Governing Board consisting of 18 members plus the chairman. The members of the Board are elected by the members of CCR. Those who have served as chairmen of the board are shown in Table I.

Chairmen are elected by the Board and alternate between industry and academia. Board membership is distributed evenly between industry and academia. To the extent possible a balance between chemistry and chemical engineering is maintained.

Table I. Chairmen of the Governing Board of the Council for Chemical Research

M. E. Pruitt, Dow Chemical, 1981, 1982
A. L. Kwiram, University of Washington, 1983
W. J. Porter, Jr., EXXON, 1984
K. B. Bischoff, University of Delaware, 1985
K. I. Mai, Shell, 1986
P. G. Gassman, University of Minnesota, 1987
E. C. Galloway, Stauffer Chemical, 1988
C. J. King, University of California, Berkeley, 1989
H. S. Eleuterio, duPont, 1990
J. I. Legg, Auburn University, 1991
R. D. Gerard, Shell, 1992
T. F. Edgar, University of Texas, Austin, 1993

CCR's programs and activities are designed to improve research communication, enhance the quality of science education, and encourage a rational discourse on the impact of science and technology on society. The programs can be broken down into seven major areas as summarized in Table II.

Table II. Council for Chemical Research Programs with Responsible Committees

Science and technology transfer meetings. (University/Industry Interaction Committee)

Programs to address improvement of science education and the supply of scientists and technologically trained people. (Public Relations and Scientific Manpower and Resource Committees)

Production and distribution of video tapes on the impact of science and technology. (Public Relations Committee)

Testimony to Congressional committees and recommendations for key government positions in science and technology. (Government Relations Committee)

Annual meeting. (Program Committee)

Awards. (Awards Committee)

Distribution of unrestricted support for universities. (Chemical Sciences & Engineering Fund Committee)

Science and technology transfer meetings assume two formats which in many respects are mirror images of each other. One type of meeting involves identification by leading industrial scientists of basic research needs for emerging new industrial technologies. Two of these so called CCR University/Industry Interface Symposia are summarized in Table III.

Table III. Industry to University Transfer: CCR University/Industry Interface Symposia

"Opportunities for Basic Research in Industrial Separations" AIChE Meeting, San Francisco, November 1989

"Basic Research Needs for Tomorrow's Industrial Catalysis" 12th North American Meeting of the Catalysis Society, Lexington, Kentucky, May 1991

The emphasis in these symposia is on transfer from industry to academia. Industrial scientists make presentations to a primarily academic audience with the objective of motivating scientists working in basic research to attack problems of interest to industrial researchers.

James Roth (Air Products), a member of CCR's Governing Board and a leading scientist in catalysis research, noted in a recent C&EN article (September 3, 1990, p. 30) that our nation is lagging behind Japan in catalysis development. CCR's participation in the conference on catalysis to be held in Lexington in 1991 is in response to this concern.

The mirror image of these meetings are the so called NICHE conferences summarized in Table IV. NICHE is the acronym for New Industrial Chemistry and Engineering Conferences.

Table IV. University to Industry: New Industrial Chemistry and Engineering (NICHE) Conferences

"Future Directions in Polymer Science and Technology." Keystone Resort, Colorado, May, 1990.

"Future Directions in Environmental Science and Technology." Scheduled for Spring 1992.

The emphasis in these conferences is on transfer from academia to industry. A recent highly successful NICHE conference on polymer science and technology was held at the Keystone Resort in Colorado. Key to the success of the conference was the organizing committee consisting of industrial representatives from major U.S. companies, namely, Air Products, Exxon, duPont, General Electric, Union Carbide, Eastman Kodak, and Dow. The organizing committee, chaired by Lloyd Robeson (Air Products), invited key polymer scientists from academic institutions in the U.S., Canada, France, and Japan to make presentations to a primarily industrial audience.

The meeting followed a Gordon Conference format with extensive opportunity for informal contact. Attendance was limited to 100. A survey of attendees showed that over 90% found the conference valuable and encouraged continuation of this meeting format. A NICHE conference on "Future Directions in Environmental Science and Technology" is now on the drawing boards. Robert Moolenaar (Dow) is chairing the organizing committee.

Because of our unique industrial/academic interface, we are in a strong position to support programs concerned with the improvement of science education and the supply of scientists and technologically trained people. Several examples of our activities in this area are shown in Table V.

Table V. Education Programs Supported by the Council for Chemical Research

Seminal workshop for middle school science education sponsored with EXXON.
Financial and advisory support for the production of the "World of Chemistry," an Annenberg/CPB Project educational television series.
Support for Operation Progress directed by Glenn Crosby at the 11th Biennial Conference on Chemical Education.

The exploratory workshop held in February, 1990, at the EXXON Education Center in New Jersey, focused on programs that are being developed by industry and academia to improve middle school science education. As a result of this seminal workshop additional workshops are being held. Information from these workshops will be used in an intersociety project initiated by Paul Gassman, Past President of the American Chemical Society, to produce and disseminate a comprehensive survey of education programs.

The Council for Chemical Research played a critical role in the production of the World of Chemistry. CCR was responsible for raising most of the industrial support needed for the project. In addition, I represented CCR on the World of Chemistry Advisory Committee where I helped edit the series.

The industrial support, together with support from the Annenberg/CPB Project and the National Science Foundation, was used to produce 26 half hour video tapes for an educational television series for college and high school students. The series is also serving an important function in conveying the central role played by the chemical sciences and technology in our society.

Dow Chemical USA and Exxon are utilizing the World of Chemistry tapes in employee orientation programs, and Dow is expanding usage overseas. The tapes are finding their way into precollege classrooms, for both student education and teacher enhancement. The World of Chemistry had the largest prebroadcast distribution of any educational television series in history. The series is being aired by Public Television to affiliates across the country.

The Council for Chemical Research is currently involved in helping raise support for The Molecular World, a prime time public television series similar in concept to the Planet Earth series. Nobel Laureate, Roald Hoffmann, the narrator in the World of Chemistry series, is the Science Editor for The Molecular World.

As a final example of our involvement in educational programs, CCR provided Glenn Crosby with partial support for the highly successful Operation Progress program for high school chemistry teachers in the southeast held in conjunction with the 11th Biennial Conference on Chemical Education in Atlanta in August, 1990.

Fred Leavitt, Executive Director of CCR, and I were witness to the Operation Progress workshop's success. The workshop included experience with computers, lectures on teaching chemistry by Professor Crosby, and extensive hands on chemistry focusing on micro laboratory instruction and a fascinating laboratory built completely around materials that can be purchased in local grocery and hardware stores. This creative laboratory when used by students and teachers should not only promote an interest in chemistry but underline the pervasiveness of chemistry by connecting the laboratory directly to the real world.

Science and mathematics education is closely related to human resources. NSF's recent restructuring of the education directorate to include human resources underlines this vital link. However, the question of human resources is complex. There are those who believe we are in a crisis situation and others who believe that supply and demand will take care of our needs. It is important that we do not give a mixed message to the United States Congress in our appeals for support for research and education.

CCR's Scientific Manpower and Resources Committee held a workshop on human resources in October, 1990. Representatives from NSF, AIChE, and ACS participated. This is a first step in our effort to bring about a more cohesive understanding of the manpower issue and, thereby, facilitate finding effective solutions to the problem.

The Council for Chemical Research has also been involved in the production and distribution of video tapes designed to educate scientists, students, educators, government scientific decision makers, and the public about the impact of chemicals on the environment. These tapes are listed in Table VI.

Table VI. Video Tapes Produced and Distributed by the Council for Chemical Research

"Carcinogens, Anticarcinogens and Risk Assessment" Presented by Bruce N. Ames, University of California, Berkeley.
"Changing Patterns of Cancer in the United States." Presented by Philip S. Cole, University of Alabama Medical School.
"Big Fears . . . Little Risks" Narrated by Walter Cronkite. Sponsored by the American Council on Science and Health.

Two tapes have been made in an attempt to put into rational perspective public fears of man made chemicals in the environment. One tape involved a lecture by Bruce Ames, an expert on risk assessment, and the other a lecture and interview with Philip Cole, one of the world's leading cancer epidemiologists. CCR has also supported the distribution of a tape entitled, "Big Fears . . . Little Risks" narrated by Walter Cronkite which does a very nice job of bringing together the material presented by Professors Ames and Cole.

The focal point of CCR's activities is the Fall annual meeting. Meetings have been held across the United States for the past 10 years, hosted by local member industries and universities. The annual meeting provides an opportunity for the entire membership to interact in programs focusing on areas of mutual interest to university, industrial, and government scientific decision makers.

The 1989 annual meeting had education as one of its themes. The 1990 meeting had programs dealing with environmental control in the '90's, frontiers in chemical sciences and technology, and university/industrial/government interactions in research.

During our annual meeting, CCR makes an award to a person singled out as most exemplary of CCR's mission. The "Mac Pruitt Award" recognizes outstanding contributions to the progress of chemistry and chemical engineering by promotion of mutually beneficial interactions between universities and chemical industry. The award includes $5000 to be given by the awardee to a chemistry or chemical engineering department of his or her choice.

Although I have stressed university/industrial interactions in this talk, I should note again that CCR's symbol contains three interlocking components, one of which is government. In order to enhance CCR's dialogue with government, it was decided to move the Bethlehem, Pennsylvania, headquarters to Washington, DC, in 1990. CCR is involved in testimony to Congress and recommendations for key government posts in science and technology. We are working with other scientific organizations to improve the effectiveness of these activities.

As noted earlier in the paper, "Although CCR cannot claim to be a primary player in helping maintain the health of a vast and far reaching industry, it has had an impact." The interaction between industry and academia that has developed through the activities discussed has been a catalyst to improved interaction on the one-on-one level. Many of us can attest to the exchange of research and technology information through visits to each other's research operations brought about because of initial CCR contacts. These visits have led to funding commitments to universities by industry. Financial gains demonstrating the increased investment of industry in academic basic research since CCR was formed, are shown in Table VII.

Table VII. External Funding for Academic Research

	Chemistry		Chemical Engineering	
	Total Funding	Industrial Funding	Total Funding	Industrial Funding
	($M)	(percent of total)	($M)	(percent of total)
1981	186	7	52	23
1983	278	9	90	31
1985	345	11	91	44
1987	500	12	114	42

A recent article in Chemical and Engineering News (November 5, 1990, p. 24) on the NSF Presidential Young Investigator Program had the following to say about CCR:

> Chemists and chemical engineers have done somewhat better than average in obtaining industrial funds, in no small part because of the efforts of the Council for Chemical Research . . . . The organization itself has actively sought to spread the word about the PYIs by distributing information to its members about each new group of awardees.

The quote refers to the behind-the-scenes work of CCR's University/Industry Interaction Committee.

Although industry targets most of its support of academic research, an industrially supported Chemical Sciences and Engineering Fund has been established by CCR that provides small amounts of unrestricted support for chemistry and chemical engineering departments. As those in academia can appreciate, unrestricted support is often of considerable value.

The Council for Chemical Research's potential for further contributions is significant. There is still much that can be done to improve academic/industrial interactions. In the Spring 1990 issue of Issues in Science and

Technology, Paul Gray, President of MIT, wrote an article entitled "Advantageous Liaisons" in which he discusses the different mode of academic/industrial interaction practiced by foreign companies. Gray states:

> "The Japanese in particular have proven themselves adept at scouring the world's pool of new knowledge, objectively exploring and evaluating the potential of novel ideas, and boldly investing in the ones with the most promise. As part of this process, they have sought to establish ties to many American research universities. They join liaison programs, they sponsor on-campus research, they send visiting scientists to university laboratories, and they even provide funding for university endowments, helping to maintain the long-term health and productivity of the resource."
>
> "In contrast, many U.S. firms have been much more reluctant to invest in the longer-term potential of emerging technologies. And they are much less willing to seek out and pursue ideas that come from outside their own laboratories, even from U.S. universities. Many faculty members are complaining these days that whereas U.S. firms have resisted their pleas to support novel research programs on campus, Japanese firms are often standing in line."

In the June 25, 1990, Wall Street Journal, Alan Murray and Urban Lehner, in a cover story article entitled, "Strained Alliance: What U.S. Scientists Discover, the Japanese Convert--Into Profit," support Gray's thesis but go beyond it. They recognize that improving the academic/industrial interface is a two way street. In support they quote a recent Ph.D. graduate in engineering as follows:

> In graduate school in the United States, everybody wants to be in basic research, not commercial research. It's more aesthetically pleasing. Graduate students believe that if you join a company, your work may not be published and you won't advance as a scientist.

As noted by Murray and Lehner, "In applying technology, U.S. falls short for reasons both corporate and cultural." The issue is clearly complex and has no simple solution. Indeed, a growing number of U.S. industries think of themselves as global companies.

CCR has a Science Advisory Board whose assignment is to look to the future with the directive to recommend programs that best take advantage of CCR's unique industrial/academic interface. The Science Advisory Board is currently looking into the relationship between CCR and the growing international environment in which chemical research and development is being conducted.

In the Wall Street Journal article Murray and Lehner conclude with a discussion of the inroads Japan is beginning to make in the aerospace industry. They, again, note that these inroads are being made by using technology that was created in the U.S. but largely left unexploited. Lawrence Clarkson, Senior Vice President at Boeing, states of the Japanese, "Today they're looked at by the world as subcontractors. Fifteen years from now they'll be major players . . . ." Iida, chairman of Mitsubishi, is more pointed when he say, "If you (the U.S) get out of the business of making airliners, if you let what happened to you in other industries happen in aircraft, your economy will really go downhill."

Is this an exaggeration? And most certainly it could never happen in chemistry and biotechnology. But there is Germany with the largest chemical companies in the world and then there is East Germany . . . or there was East Germany.

If we are to maintain a healthy economy, we are going to have to continue to improve industrial/academic ties. We still have the world's greatest universities and we are still leaders in the chemically based industries. We must not only maintain this leadership, we can strengthen it . . . if we work together.

The Council for Chemical Research is an organization with growing recognition for its unique structure and potential. We are in a position to achieve what other organizations may not be able to accomplish. Within the last year, CCR has been approached by the Dreyfus Foundation, Research Corporation, and the National Science Foundation to consider joint efforts between these organizations and CCR in which CCR will play a central role. Our unique industrial/academic interface is emphasized in these solicitations.

Henry McGee, Director of the Division for Chemical and Thermal Systems at NSF, in asking CCR to play a major role in developing an industrial/NSF supported program for single investigators, wrote:

> "A major outcome of CCR has been the developing trust and rapport between industrial and academic practitioners of pure and applied chemistry. Your organization is well positioned to assist me."

It is precisely this trust that is at the core of CCR's success. The trust has evolved as a result of the personal relationships that have developed through the years between our industrial and academic members who have worked together on our many projects. The wide spread mistrust that permeated the first meeting in Midland eleven years ago is being replaced by the trust factor.

The potential of the trust factor has just been tested. In many ways the Council for Chemical Research is still in its infancy, still an experiment. CCR's full impact is still to be realized. If we are to effectively meet the challenge of remaining, as Philip Abelson puts it, "tops in chemistry," we need to continue to build on the trust factor. The uniqueness of CCR's structure and the focus of its mission places the Council for Chemical Research in a singular position among scientific organizations.

Yes, it is possible for industry and academia to work together to improve the welfare of this country, and there is much left to do. Perhaps CCR's major long term impact will be that it will serve as a model for others to follow.

RECEIVED May 31, 1991

Chapter 3

Undergraduate Research

Academic–Industrial Partnerships

Michael P. Doyle

Department of Chemistry, Trinity University, San Antonio, TX 78212

The Council on Undergraduate Research, a Society for the Advancement of Scientific Research at Primarily Undergraduate Colleges and Universities, initiated a new program in 1990, sponsored by science industry and designed to attract talented undergraduate students into careers in science. In this program 20 students from predominantly undergraduate institutions were selected to receive $2500 fellowships to conduct research full time, normally during the summer months, with their faculty mentor. For the summer preceding their entrance into graduate school, industrial sponsors offer their awardees the opportunity to work in their corporate research laboratories. These Research Partnerships link student and faculty member with the fellowship sponsor to provide them with an association for mutual identification not otherwise possible. Sponsors in 1990 were American Cyanamid Company (Agricultural Research Division), Eli Lilly and Company, Hewlett-Packard Company, the Merck Company Foundation, Norwich Eaton Pharmaceuticals, Inc., Pfizer Central Research, Rohm and Haas Company, and SmithKline Beecham Pharmaceuticals.

From a high of nearly 12,000 in 1978, the number of students who graduate from U.S. colleges and universities with a degree in chemistry has dropped precipitously during the 1980's (Figure 1). For 1989, the last year for which we have this information, this number was 8,122 (*1*) and further decreases can be expected during the early 1990's. A loss of chemistry majors was not unanticipated, because during this same period the United States experienced a decline of approximately 20 percent in the number of 22-year olds, but the decrease in the number of chemistry graduates exceeds by nearly 50 percent that anticipated from the demographic pool.

Student interest in the physical sciences has been declining since the mid-1960's. A study conducted by the American Council on Education and the Department of Education at UCLA has for many years surveyed the interests of entering college freshmen (*2*). Those planning to major in the physical sciences declined from 7.3% in 1967 to 3.8% in 1975 to 2.4% in 1983. In absolute

0097–6156/92/0478–0016$06.00/0

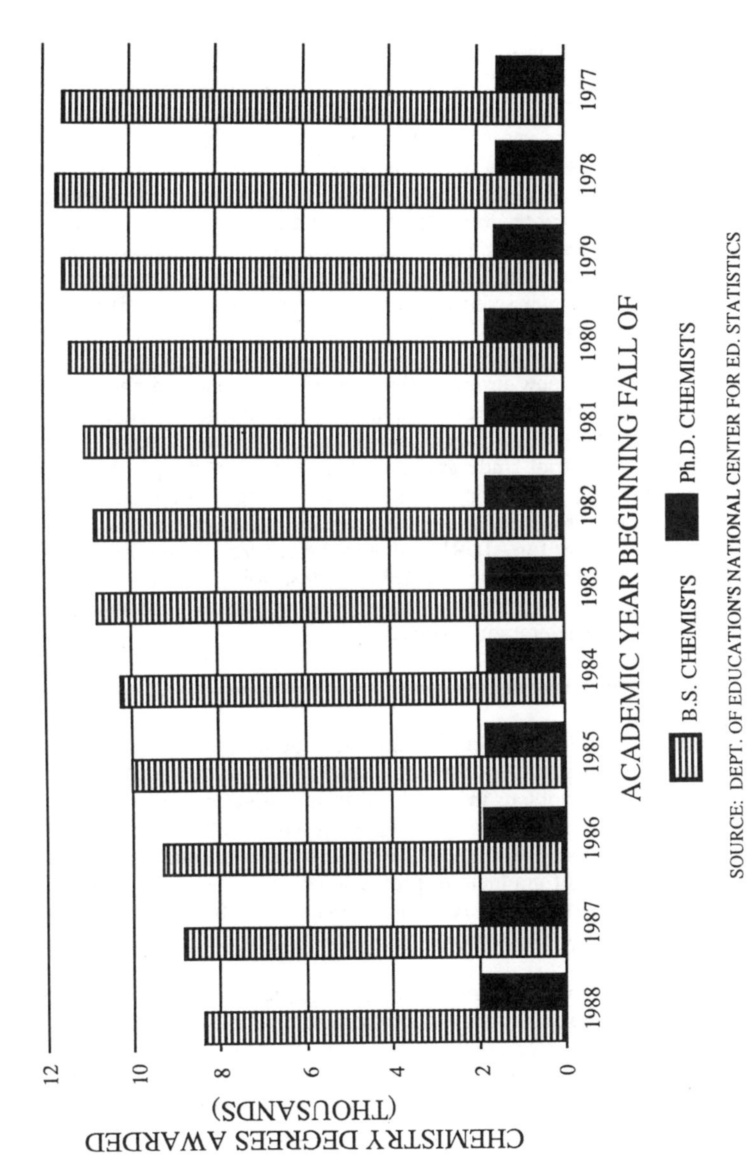

Figure 1. Supply of new chemists. Data for 1986-1989 have been obtained from the American Chemical Society Committee on Professional Training (*1*). Prior years are those from the Department of Education's National Center for Education Statistics.

numbers, these percentages represent 96,400, 58,700, and 37,600 students for the respective years and an overall decline of 61% in only 16 years.

What is perhaps of equal concern is the impression that the quality of students who are entering the chemical sciences has also declined. The first reference to this phenomenon was that from a controversial report produced by Douglas Neckers of Bowling Green State University in 1979 from a study funded by the Sloan Foundation (*3*). Neckers examined SAT scores of graduating chemistry majors against their career plans for the period 1966-1978, and he concluded that the capabilities and qualifications of students who pursued advanced degrees in chemistry declined significantly during those years. In contrast, chemistry majors choosing a career in medicine showed a substantial increase in both math and verbal SAT scores during the same period. Similar conclusions have been reached in broader student classifications by the ACE-UCLA study (2), and in the 1980's fields such as computer science and business/economics captured more talented students away from chemistry.

This chapter reports a new program that has been designed to encourage bright and motivated students to enter careers in the chemical sciences. Founded on the principle that undergraduate research is the single most important career stimulus for students, Academic-Industrial Undergraduate Research Partnerships (AIURP) offer a novel approach to the resolution of the serious shortages in the chemistry workforce. The organization responsible for this program is the Council on Undergraduate Research, and support for AIURP is due to the generosity and interest of chemical industry. However, before specifically describing AIURP and its operation, the following sections will first describe the origins of undergraduate research and the Council on Undergraduate Research.

The Origin of Undergraduate Research

Undergraduate research is a relatively new educational venture and, in all respects, is an American invention. Growing out of the fertile combination of an investigator in search of a problem with a problem in search of an investigator, undergraduate research has become the most exciting educational development of the second half of the 20th century. Its success is measured, in part, by the stimulation of this experience for students to enter graduate or professional schools, but its principal benefit is that it imparts to students a realistic assessment of the character of a discipline through the process of discovery.

Like a 16-year old who has just received a driver's permit, an undergraduate student has considerable enthusiasm but lacks experience. The student may have completed most of the basic courses expected for a major in the discipline but is not yet so sophisticated to know if a question that he or she might ask has already been answered. The faculty scholar, on the other hand, is an expert in at least one area of the discipline and understands what problems are ripe for discovery. When the scholar accepts the apprentice, a problem is identified and the approach to its solution becomes the framework of an undergraduate research experience. Initially, the scholar directs all aspects of the problem's development but, eventually, the student becomes the expert.

The origin of undergraduate research is difficult to assess, and there has been different directions taken in different disciplines. In the sciences, which have the longest tradition of undergraduate research, the actual visible beginning of these experiences occurred only after the Second World War. There were, of course, examples of individuals and institutions that engaged in these activities even during the nineteenth century, but they were isolated instances peculiar to certain institutions and to teacher-scholars who promoted such experiences for highly talented students. Even Harry Holmes, a distinguished scientist and Professor of Chemistry at Oberlin College, inferred in 1924 that research was a proper engagement for the college teacher, but not necessarily for the student. In responding to an earlier criticism of college teachers who engaged in research, Professor Holmes states (*4*):

> "A stimulating freshness and a feeling of authority come to the college teacher as he unravels the secrets of science. The teacher profits, the great body of science profits, and the pupil profits. The pupil then feels that he is near one of the fresh springs that feed the stream of knowledge into which he has been dipping.
>
> It is essential that the teacher do research work, i.e., he should comb the subject of chemistry from end to end for facts and for methods of exposition that will make such facts live and real to his students."

As an educational methodology, research was to be valued because it imparted excitement into what might otherwise be an exposition of dull facts. But the involvement of unsophisticated undergraduate students in this endeavor was not expected and, for most faculty in colleges and universities, considered impossible.

Undergraduate research had its beginnings in faculty research where students took on the role of assistant, setting up experiments, preparing starting materials, or looking after experimental animals, but they did not perform the actual experiment. To do so would have led to uncertainty in the results and their interpretation, because how could an untrained eye discern the complex nature of the experiment being performed? Yet in this pre-World War II era, students were involved in many laboratories, and they were watching the conduct of experiments and learning about the process of discovery. This was especially true in undergraduate institutions where junior and senior students were the principal workforce. In universities with graduate programs there was less need to involve undergraduate students; here graduate students were available and had as their principal objective the conduct of research.

Something extraordinary occurred during this period. Undergraduate institutions educated more students who went on to obtain graduate degrees in science than did many of their larger university counterparts which had graduate school programs. In its Report to the President in 1947 on Science and Public Policy, the President's Scientific Research Board observed (*5*):

> "Although some 90 universities grant all the doctor degrees in science, undergraduate work in science is scattered throughout our higher educational system. Less than half the doctors of science receive their undergraduate training in the same school that confers their advanced

degree. The remainder complete their undergraduate work in about 600 other colleges or universities.

Thus, the 90 university graduate schools depend in large part upon 700 schools, including their own, which grant bachelor degrees in science. These in turn depend upon science courses in many others of the total of 1,700 schools in the country. Many smaller institutions have, in the past, contributed scientists out of all proportion to the numbers of their students. Thus:

During the years of 1936 to 1945, Furman University, Oberlin College, Reed College, and Miami University together graduated more students who later completed doctoral work in physics than did Ohio State University, Yale University, Stanford University, and Princeton University combined.

Over the same period, Hope College, Juniata College, Monmouth College, St. Olaf College, and Oberlin College combined produced more candidates for doctor's degree in chemistry than did Johns Hopkins University, Fordham University, Columbia University, Tulane University, and Syracuse University, all together."

Why did this occur? We recognize now that research was a significant preoccupation at these colleges, and undergraduate students observed the challenges of investigation and the enthusiasm that was generated by discovery.

Just following the end of World War II, the Research Corporation designed a funding initiative, the Cottrell Grants Program, to provide incentive for scientists to return to colleges and universities rather than joining on-going industrial and federal research at the large central laboratories into which they had been "drafted" for the course of the War (*6*). Grants from the Research Corporation were provided to faculty for research in which they were engaged or about to initiate, and a significant fraction of these grants were awarded to scientists at predominantly undergraduate institutions. Not surprisingly, in those early years most of these Cottrell grants were received by faculty at institutions that already had a recognized tradition of research. But these grants, unlike contracts provided by the Office of Naval Research (ONR) at that time, made possible full time summer research for selected undergraduate students.

Faculty members who were performing research in the sciences during the summer needed assistants and, without the cadre of free labor available when classes were in session, found that student employment provided the necessary workforce. Funding available from the Research Corporation provided this flexibility. No longer limited by the time constraints of coursework, students become more intimately involved in actual experimentation. They learned the techniques and mastered observation with critical evaluation of results. By the end of the summer, these summer employees were well versed in experimental details and filled with the excitement of potential new discoveries. With the advent of the new academic year, faculty curtailed their research in order to prepare for classes, but their students, enthusiastic with experience from their summer research engagements, came into the laboratory with regularity to continue their experimentation. Out of this was borne the beginnings of undergraduate research in the sciences.

The next major leap in the development of undergraduate research occurred in the early 1960's when the National Science Foundation, which had been created by Congress only a decade earlier, initiated the Undergraduate Science Education program which became their Undergraduate Research Participation (URP) program. In this post-Sputnik era, this country placed a high premium on encouraging students into careers in science and engineering, and the URP program was created with the belief that if you allow an undergraduate students to experience the challenges and excitement of discovery, their participation would become an addiction.

The success of the URP initiative during the 1960's can be measured in terms of the rapid increase in the numbers of students who obtained their Ph.D. degrees in the sciences (*7*) and in the comments of URP students who found that their undergraduate research experiences led them to careers in the sciences (*8*). The URP program had its greatest impact on students in biology, chemistry, and physics - so much so that by the early 1970's more students obtained their Ph.D. degrees in these fields than there were positions available to them. Grants awarded to public and private colleges and universities opened new vistas for many institutions without prior experiences in undergraduate research and, in many respects, the enterprise was institutionalized in the sciences during this period.

Although the largest single contributor to the development of the tradition of undergraduate research in the sciences, the National Science Foundation's Undergraduate Research Program was not the only initiative. Research grants to faculty awarded by the National Science Foundation and the National Institutes of Health were often used, in part, to support undergraduate research. The Petroleum Research Fund, administered by the American Chemical Society, encouraged undergraduate research through their Type B grants to faculty in undergraduate departments at colleges and universities. The Research Corporation continued its funding ventures and, in 1971, initiated its College Cottrell Science Program to support faculty and student research at private (now public and private) undergraduate institutions. Even organizations as diverse as the Argonne National Laboratory and Du Pont hired undergraduate students to undertake research experiences in their laboratories. Similar support mechanisms for undergraduate research did not exist in the social sciences, humanities, and the arts.

In part because of the excess supply of scientists in the early 1970's, various attempts were made to dismantle the URP program, but without success. Instead, its goals and targets were changed from year to year until 1981 when this program, along with the entire science education operation at the NSF, was terminated. Unfortunately, the URP program was lost at the same time that the number of new Ph.D.'s entering the mainstream of science was declining to pre-1965 levels. Reconsideration of this impact, principally through a comprehensive study of undergraduate science, mathematics, and engineering education by a Task Force of the National Science Board (*9*), as well as efforts undertaken through the NSF's Chemistry Division, resulted in the resurrection of undergraduate research participation through introduction of the NSF's Research Experiences for Undergraduates (REU) program, now in its fifth year. Other efforts to stimulate an increase in the interest of talented students for science and to enhance the tradition of research in undergraduate institutions, including the Mentor/Scholar program of the Camille and Henry Dreyfus Foundation, the private college/university consortial science programs funded by the PEW Charitable Trusts, and the Hughes bioscience initiative have also been recently introduced, but only the Dreyfus Foundation program appears to have continuity.

The very nature of undergraduate research requires a special talent in the preceptor. The problems undertaken must be significant but they must also be doable within a limited time frame, and students must be given the opportunity to develop the investigation. Often the research begun by one student is continued by another. In other approaches teams of students are engaged, each assigned to a particular aspect of a problem, or the preceptor and student approach the investigation together, each contributing to its development. No single model is appropriate to all investigators or all investigations.

Twenty years ago undergraduate research was limited in most institutions to students in their senior year, and the term "senior research" was commonly applied to this endeavor. The remnants of this are still seen in "senior honors projects" at many colleges and universities. However, such limitations may actually inhibit the development of students in a research program since their graduation abruptly terminates their investigations just when they are most capable of obtaining critical results. Instead, early entry into research allows students the luxury of learning about research, making mistakes, and understanding pertinent literature with time remaining to thoroughly investigate the problem.

The Council on Undergraduate Research

The Council on Undergraduate Research (CUR) is a society for the advancement of scientific research at primarily undergraduate colleges and universities. Founded in 1978 by a group of liberal arts college chemistry faculty, CUR has expanded to serve all the sciences and mathematics at undergraduate colleges and universities. The purposes of this Society are to provide undergraduate students at these institutions with increased opportunities to learn science by doing it and to provide their science faculty with increased opportunities to continue to develop their own understanding of science by remaining active in research. CUR believes that a discovery-oriented approach to learning should permeate science education throughout the undergraduate science curriculum. As described in the previous section, increased opportunities for students to do research as undergraduates effectively draw more students to careers in science teaching and research, and continuing involvement in research assists faculty to become more exciting and stimulating classroom teachers.

As emphasized by Williams College President Francis Oakley in his recent address at a National CUR Conference (*10*), the diversity and comprehensiveness of the American system of higher education are unparalleled by any society in any era. At one end of the spectrum are the great research universities, where faculty research is so important that it sometimes overshadows undergraduate teaching, but where faculty expertise, facilities, and equipment are readily available to support research by interested undergraduates. At the other end of the spectrum are institutions where limited resources preclude research by students or faculty. The majority of American institutions of higher education lie between these two extremes. With encouragement, with sharing of successful models, with modest local resources, and with help obtaining external support, faculty at these middle range institutions can develop programs that introduce students to the excitement and challenge of science by doing research as undergraduates without first plowing through four years of traditional lectures, laboratories and problem sets.

The accomplishments of the Council on Undergraduate Research as a grass-roots movement have been substantial:

- CUR publishes directories which document the very significant role of undergraduate departments and their faculty in the mainstream of science. As a result, funding agencies use the directories in the evaluation of proposals and selection of reviewers. Graduate schools use the the directories in their recruitment efforts, companies use them in the search for talented graduates, and they are even used by some high school students in selecting colleges. Currently, there are directories in biology (Second Edition, 1989, 618 pages, 89 institutions), chemistry (Fourth Edition, 1990, 747 pages, 226 institutions), geology (First Edition, 1989, 682 pages, 133 institutions), and physics/astronomy (Second Edition, 1989, 537 pages, 124 institutions). The first directory for mathematical sciences, which established a Council in 1989, is in preparation. Initial support for the chemistry directory was provided through a grant from the Petroleum Research Fund of the American Chemical Society.

- CUR publishes a *Newsletter* in four 100-page issues annually to provide members of CUR and non-member subscribers with successful models for research programs and for their support through acquisition of outside funding. The experiences of CUR members and others in designing and implementing programs in response to special foundation initiatives are disseminated. The *Newsletter* pays special attention to sources of funding, including the names and telephone numbers of persons to contact for information. Now in its eleventh year, the *Newsletter* is distributed to more than 1350 individuals.

- Biannually, CUR sponsors a national conference to examine critical issues affecting science education at primarily undergraduate institutions. The third such conference brought nearly 300 science faculty, college administrators and representatives of federal agencies and private foundations to Trinity University in San Antonio in June, 1990, to examine "The Role of Undergraduate Research in Science Education: Building and Funding a Successful Program". Networking among college scientists involved in other cooperative efforts to enhance undergraduate science education is a very important aspect of these conferences and meetings. For example, the National Conferences on Undergraduate Research (NCUR), organized separately from CUR and accepting papers for presentation at each annual conference from students in all academic disciplines from all colleges and universities, were conceived and first implemented by a CUR councilor, and several CUR councilors currently serve on the NCUR Board.

- In 1989 with support from the Research Corporation, CUR instituted a consulting service to advise chemistry departments about ways to improve their programs and increase their success rate in obtaining external grants. The program includes a visit to the department by two CUR consultants, who meet with faculty, students, and administrators and who later submit written recommendations. Followup visits to the department by the consultants and by department members to the consultants' institutions are encouraged.

- Beginning in the summer of 1990, CUR has offered to selected students Academic-Industrial Undergraduate Research Partnership (AIURP) fellowships in cooperation with leading American scientific companies. These fellowships provide $2500 to students to allow them to engage in research with faculty mentors at their home institutions normally during the summer after their junior year and, with most industries, provide these same students with the opportunity to work in the industrial sponsor's research laboratories during the summer preceding their entrance into graduate school.

- In 1983, CUR submitted a proposal to the National Science Board that was implemented as the NSF Research in Undergraduate Institutions (RUI) initiative (1984). After its first year the RUI program was reviewed by an ad hoc group that included among its four faculty members two chemists who were CUR councilors and a physicist who was to become a CUR councilor. The RUI program has become the model for "distributed funding" of science education through the NSF research directorates.

- Other CUR efforts to stimulate government interest in funding science at undergraduate institutions have included involvement with the development of the NSF Instrumentation and Laboratory Improvement (ILI) program, the NSF Research Experiences for Undergraduates (REU) program, and the NIH AREA program. CUR councilors helped to plan and chair sessions at the AREA workshop held in Bethesda, MD in March, 1990 (*11*).

- The visibility of CUR to agencies and foundations has led to increased representation by undergraduate institution science faculty on important policy-making and funding committees. These have included advisory committees and review panels for the National Science Foundation, panel members for the National Institutes of Health, membership on the National Research Council's Board on Chemical Sciences and Technology, membership on advisory panels for private foundations, and service on boards of foundations and other scientific societies. The changes in these activities over the 12-year history of CUR have been enormous.

Initially formed in 1978 by chemistry faculty at private liberal arts colleges, CUR expanded to include public and private colleges and universities in 1983 and to include additional disciplinary councils in physics/astronomy and biology in 1985, geology in 1987, and the mathematical sciences in 1989. Prior to June 1989 the Council on Undergraduate Research consisted solely of councilors elected from among their colleagues by the current councilors, and all of its operations have been voluntary. Committees were staffed by volunteers from among the councilors for the preparation and publication of the CUR directories and its Newsletter, for the arrangements and planning for National CUR Conferences, and for other assignments approved by the Executive Committee or the full Council.

In response to faculty and administrators across the country who expressed interest in getting involved in CUR and in order to provide a larger and more open forum for discussions of issues, CUR began in September 1989 to enroll members, who in turn elect councilors from within the membership. During the first year more than 1200 applications for membership were received, including blocks of applications from single institutions numbering as high as 85. A full-time Executive Officer will be selected in 1991 with support received through grants received from the PEW Charitable Trusts and the Research Corporation and with contributions from undergraduate colleges and universities.

Academic Industrial Undergraduate Research Partnerships

With undergraduate research programs reinstituted into colleges and universities, and with expanded opportunities offered by National Laboratories, the one venture that had not been undertaken, except with only a handful of chemical companies as more than a local initiative, was student involvement in industrial research. The vast majority of students drawn into chemistry careers enter industrial positions, yet very few of them ever have the opportunity to experience these environments prior to entering full-time employment.

The experience of one of my research students, Wendell Wierenga, taught me the motivational value of research in industry for career development. In the late 1960's, Du Pont initiated a program in which one student from each of several selected undergraduate institutions was invited to spend the summer following graduation at the Experimental Station. Wendell, who was to co-author seven research publications with me from only one year of undergraduate participation, was selected by the chemistry department. His experience at Du Pont led him to consider industry favorably when he had to select from among academic and industrial positions following the award of his Ph.D. He chose employment at Upjohn and initiated a similar undergraduate research program that has allowed Hope College and, subsequently, Trinity University students, among others, to work at Upjohn either before or after their senior year.

Many of the most talented students from my group have benefited from their experience in industry, and all who have had this opportunity have been favorably impressed with the challenges, environment, and rewards that they have encountered. More recently, two additional programs have been initiated by former students at Exxon in Baton Rouge (Bruce Cook) and at Norwich Eaton in New York (Charles McOsker).

The reason for the success of these programs lies in the early experience of undergraduate students at their home institutions. Research participation coupled with motivation from their mentors helps these students to decide their career direction and their relative capabilities for discovery. They begin their industrial experience well prepared in laboratory techniques and with instrumentation, and questions that have fermented in their minds about relevance and opportunities -

questions not easily answered in an undergraduate environment - can be addressed in the industrial setting.

These considerations have led the Council on Undergraduate Research to introduce Academic-Industrial Undergraduate Research Partnerships (AIURP) - a program designed to broaden research opportunities for selected undergraduate students with a strong linkage to chemical industry sponsors. In this program competitive fellowships are made available to highly talented and motivated students at undergraduate colleges and universities to engage in research at their home institutions, ordinarily during the summer following their junior year. Opportunity is then provided for these same students to be employed in the industrial research environment of their sponsor during the summer before they enter graduate school. Industrial sponsors provide the initial fellowship award of $2,500 per student per summer and, if acceptable in theory and feasible in practice, they make available for their AIURP fellowship recipient a summer research position during the following summer in their own laboratories. The Council on Undergraduate Research advertises this program, reviews applications and selects students, and links the fellowship awardee to the industrial sponsor.

In 1990 twenty AIURP fellowships were provided by eight industrial sponsors:

American Cyanamid Company, Agricultural Research Division
Eli Lilly and Company
Hewlett-Packard Company
Merck Company Foundation
Norwich Eaton Pharmaceuticals, Inc.
Pfizer Central Research
Rohm and Haas Company
SmithKline Beecham Pharmaceuticals

Three quarters of these fellowships were awarded to chemistry students and the remainder to biology and physics students. The Monsanto Company has joined this list of sponsors for 1991, and other companies and organizations are considering this program. The design of the AIURP program, whose fellowships were offered for the first time in 1990, implemented suggestions made by several sponsors, particularly Merck and Rohm & Haas. Applicable disciplines are determined by the interests and specifications of AIURP sponsors.

Eligible students are those who plan to enter graduate school in science or engineering, have a 3.3 or higher grade point average, and who normally will have completed their junior year of studies. The quality of the research in which the student is to be engaged, the qualifications of the faculty mentor, and the facilities available for the conduct of this research are considered in making selections.

Student response to the first year of AIURP awards has been overwhelmingly positive. About one of the participants, her mentor wrote "her attitude toward research is one of her greatest attributes. She comes in early in the morning and is prepared to work. She takes her notebook (a very neat notebook) home with her at night so she can plan the reactions for the following day. I think that she will be an outstanding researcher in the future." About her summer research experience, one AIURP fellowship recipient wrote: "The summer of 1990 was one of tremendous growth both for me and the lab in which I worked . . . I was in a unique position to assist in laying the ground work for years of rewarding research . . . In short, this summer has changed me from a one-dimensional book-learner into a three-dimensional scientist with the skills and perspective to make my own contribution to the body of knowledge I once felt imutable." These students will apply to graduate school and, following experience in their sponsor's laboratory during the summer of 1991, each should be poised to undertake their studies with renewed enthusiasm.

Literature Cited

1. *Chemical & Engineering News* **1990**, *68(18)*, 29.

2. Solomon, L.C.; La Porte, M.A. *J. Higher Educ.* **1986**, *57*, 370.

3. Neckers, D.C. *On Chemistry Majors, 1961-1979* Bowling Green State University: Bowling Green, OH., 1981.

4. Holmes, H.N. *J. Chem. Educ.* **1924**, *1*, 81.

5. The President's Scientific Research Board *Science and Public Policy: A Report to the President*; Steelman, J.R., Ed.; August 27, 1947; pp. 19-21.

6. Schauer, C.H. *CUR Newsletter* **1982**, *2(2)*, 32.

7. Coyle, S.L.; Bae, Y. Summary Report 1986: Doctorate Recipients from United States Universities; National Academy Press: Washington, D.C., 1987.

8. Hayford, E.R.; Salda, M.L.; Reif, K. *CUR Newsletter* **1986**, *7(1)*, 35.

9. Neal, H.A. *Undergraduate Science, Mathematics, and Engineering Education*; National Science Board: Washington, D.C., 1986.

10. Oakley, F. *CUR Newsletter* **1990**, *11(1)*, 15.

11. Mohrig, J.; Lieberman, E.C. *CUR Newsletter* **1990**, *10(4)*, 57.

RECEIVED May 31, 1991

Chapter 4

Ohio's Thomas Edison Program

A Technology Transfer Model

E. C. Galloway

Edison Polymer Innovation Corporation, 3505 E. Royalton Road, Broadview Heights, OH 44147

The subject "Technology Transfer" has been treated in numerous articles in recent years, with the point usually being that we don't manage it well in the U.S. While our universities are the best in the world in research, we are slow to capitalize commercially on the results. Frequently, our overseas competitors take our research results and transform them into products, processes and jobs more effectively than we do. In the 1980's, several states established technology transfer programs to address this problem. In particular, those states which were hardest hit during the recession years recognized investment in technology as an appealing alternative to investment in mature industries, and each state gave serious thought to its program. The programs were structured deliberately to match perceived needs and overall economic strategies. While the state programs differed in structure and emphasis, the desired end result was always the same: diversification and economic growth through the creation of new products, manufacturing facilities and jobs, and by strengthening existing industries.

For state governments, providing support of R&D as a means of bringing about economic development is a relatively new concept, but one which has become increasingly necessary in order to maintain a competitive industrial base in a rapidly changing, technology oriented global economy.

The studies of Mansfield (*1*) and others clearly establish that investment in R&D produces economic advancement. Corporations that invest in R&D perform better over the long-term, with new products and processes, and with better growth in sales and earnings. At the macro level, national economies benefit from reduced unemployment, advances in public health, a stronger defense capability and general improvement in the quality of life.

In 1984, the State of Ohio launched the Thomas Edison Program as a public-private partnership to promote technological innovation through university-industry linkages. The major objective of the program was to improve Ohio's economy by building on its technological strengths.

0097–6156/92/0478–0028$06.00/0

This chapter will review state technology transfer initiatives, Ohio's Thomas Edison program, and the policies and programs of The Edison Polymer Innovation Corporation (EPIC), as one of the eight Edison Technology centers.

State Initiatives: A Major Movement

Across the country, state governments have established economic development programs in many forms: science and technology centers, seed and venture capital funds, business incubators, equipment subsidies, and even endowed university chairs. In 1988, 43 states were said to have spent $550 million on such programs. These state initiatives are strategic moves based on the premise that the application of scientific knowledge can be the basis for economic expansion and diversification as new businesses are created and old ones are made more competitive.

Reflecting just how important these programs have become, the U.S. Department of Commerce has established "The Clearinghouse for State and Local Initiatives on Productivity, Technology and Innovation" (*2*) to serve as a central source of information and policy analysis.

A report from the Minnesota Governor's Office of Science and Technology (*3*) also provides a considerable amount of data on state programs. For example, according to this report, 41% of the $550 million in aggregate state funding for science and technology initiatives was for technology centers and 27% for research grants. The balance was used for research parks, incubator facilities and seed capital investment programs.

The various participants in state programs usually have very different reasons for becoming involved in them. Universities enter into relationships with government and the business community to gain additional financing for their traditional interests in conducting academic research, training graduate students, and enhancing their facilities. Corporations enter into such programs to accomplish specific research goals, find promising graduate students they would like to hire, develop long-term relationships with research professors, or to stay abreast of new ideas as they develop in the academic world. State governments have a third set of objectives: to create jobs, increase the productivity and growth of businesses in the state, and enhance the general research base within the state.

If a program's creators are not explicit about their goals, each party can operate with a very different agenda and with different assumptions, and the state's goals can easily get lost in the process. "State programs which result only in buildings and facilities for the universities and a stream of academic papers, without measurable impact on the state's economy, are failed programs (*4*)."

State programs, several of which are now more than five years old are also beginning to be judged critically by major companies, who initially may have provided funding as good corporate citizens, but now are looking for a tangible return on their research investment. However, Walter Plosila, the founding director of the Ben Franklin Partnership Program of Pennsylvania, counsels participants in these programs to remember that centers of technological innovation cannot produce results overnight. In fact they may never achieve complete financial self-sufficiency, which means that sustained public money and support may remain important for many years.

Ohio's Edison Program: A Bold Initiative

In the early 1980's, the "rustbelt" was facing a severe recession. As a result, the State of Ohio, along with Pennsylvania, Michigan and Illinois, pioneered in the creation of an economic development program based on investment in technology. Former Governor Richard F. Celeste started the Edison Program in 1984, when Ohio faced a $500 million budget deficit and the state's unemployment was running at nearly 15%.

As Christopher Coburn, former Executive Director of the Edison Program stated: "It was to be a non-traditional effort designed to support innovation, diversify the state's economy and promote entrepreneurship. It would be the perfect bridge between an enhanced university system and economic growth." Ohioans recognized that by creating the program Governor Celeste and the General Assembly were committing the State of Ohio to a broad scale support of applied research, unprecedented at that time. Said Coburn, "Departing from traditional approaches, the state pledged itself to a long-term initiative that would not pay off for many years into the future".

The Edison Program has three main components:

> (a) The Seed Development Fund that provides funds to support university-industry applied R&D activities, with grants up to $50 thousand for feasibility studies, and grants up to $250 thousand for advanced applied research.
> (b) Seven Edison Incubators, located on university campuses, that provide basic business services such as accounting, secretarial help and legal advice to new technology-based start-up companies.
> (c) Eight Technology Centers that carry out early stage generic research, perform contract research for individual clients, create technology transfer mechanisms and promote scientific education and training.

Now six years old, the Edison Program has succeeded in encouraging technological innovation, creating jobs and fostering economic growth. It has emerged as a $300 million partnership between government and industry which involves every major public and private Ohio university and more than 700 companies. Fortunately the state administration in Ohio accepts the Edison Program as a long-term initiative, and continues to give it strong support.

Edison Technology Centers

The Edison Technology centers (Figure 1) were established in specialized fields where there was both opportunity for economic growth and an established foundation upon which to build. The fields selected were manufacturing systems, biotechnology, and advanced materials, including polymers.

No single organizational model was prescribed by the state for adoption by the centers. Each center was allowed to develop independently, with its structure and program tailored to match the circumstances and requirements of its particular field.

Some centers carry out research programs within their own facilities, while others rely entirely on their university partners; some are affiliated with several Ohio universities and some with only one or two; some are regional while others are global. A complete description of the Edison Program and its several centers is provided in a recent report by the National Research Council (*5*).

The Edison Program, like most state technology programs, requires that matching funds for any research investment be obtained from industry, thus leveraging the state's investment. This serves to establish the technology transfer partnership early on and increases the likelihood that the research will match real needs or represent viable opportunities for commercialization. It also provides a check on university faculty members who might prefer to use the funds to pursue research based on more personal interests.

Small Business Program: Potential For Technology Transfer

Nationally, the small business sector contributes disproportionately to job creation and economic growth. It has been reported that enterprises with fewer than 100 employees accounted for over half of the new jobs created during the period 1980-1986, although they employed only about 35% of the total workforce. On a regional level, small companies strengthen a local economy, and the Edison Program places considerable emphasis on providing technical and financial support to such companies, as well as to entrepreneurs and start-up companies.

Studies of technological innovation have indicated that small companies are more efficient innovators than are large firms. Yet small business is the sector most starved for capital. The Edison Seed Development Fund, with its grant program, is designed particularly to provide timely assistance to these small companies.

Some of the research results from university projects are expected to lead to the spawning of new start-up companies with a large potential for business growth. The Seed Development Fund can help a start-up company arrange for the R&D necessary to obtain additional funding, enter new markets and, accelerate build-up to a relatively stable position.

The Seed Development Fund and Incubator Programs are designed particularly to help entrepreneurs and small companies, but each Edison Technology Center is also required to maintain a vigorous small member program. The centers are expected to provide assistance on problems of a technical nature, but also advise in other areas, such as locating sources of financing, satisfying government regulatory requirements or setting up a medical benefits program.

Some of the centers, such as the Cleveland Advanced Manufacturing Program (CAMP), are primarily focused on local businesses. In addition to arranging for financial assistance through such programs as the Seed Development Fund or the federal Small Business Innovation Research program, they also provide seminars, hands-on training courses and individual counseling. Through improvements in manufacturing systems, via total quality programs, statistical process control techniques and robotics, substantial improvements in productivity can be achieved. In 1989, the CAMP program was awarded a grant by the National Institute of Standards and Technology (NIST), which will provide at least $1.5 million a year for the next three years to establish one of three Manufacturing Technology Centers in the nation. The NIST center in Cleveland serves companies in other states in the midwest, as well as Ohio.

The Edison Polymer Innovation Corporation (EPIC)

EPIC was established in 1984 to take advantage of Northeast Ohio's strong research and manufacturing base in polymers. The University of Akron and Case Western Reserve University, located in Cleveland, possessed world class reputations in polymer science and engineering, with extensive facilities and large faculties. At the same time, there was a large industrial base of polymer related companies located in the region, sometimes referred to as "Polymer Valley." These included corporate giants in the plastics and rubber industries, and more than 1,000 small businesses: Suppliers of equipment and materials, and manufacturers of plastic parts and products. EPIC's role was to identify opportunities where the tremendous polymer research capability of the two universities could be brought to bear on problems and needs of the industrial sector.

Considering the barriers which may exist between technical and commercial departments within a corporation --- reflecting differences in educational background and experience, functional responsibilities and operating time frames --- one can appreciate the technology transfer challenge facing an Edison center. In practice, the initial research is carried out in an academic setting. Then, when a research project shows commercial promise, the private sector is given an opportunity to sponsor the advanced development required to move forward to commercialization. EPIC operates in the gray zone of (Figure 2), trying to facilitate the translation and movement of technology from university to company, in spite of the cultural differences.

Frederick Betz of the National Science Foundation commented on the difficulties associated with making this university-to-company transfer at a 1987 conference on research consortia (*6*): "At NSF, when we looked at the issue of why technology wasn't being transferred out of the universities to industry, the answer was --- it didn't exist in the right form, because the research wasn't done

MATERIALS

- Edison Polymer Innovation Corporation
- Edison Materials Technology Center

BIOTECHNOLOGY

- Edison Biotechnology Center
- Edison Animal Biotechnology Center

MANUFACTURING

- Institute of Advanced Manufacturing Sciences
- Edison Industrial Systems Center
- Edison Welding Institute
- Cleveland Advanced Manufacturing Program

Figure 1. Edison Technology Centers.

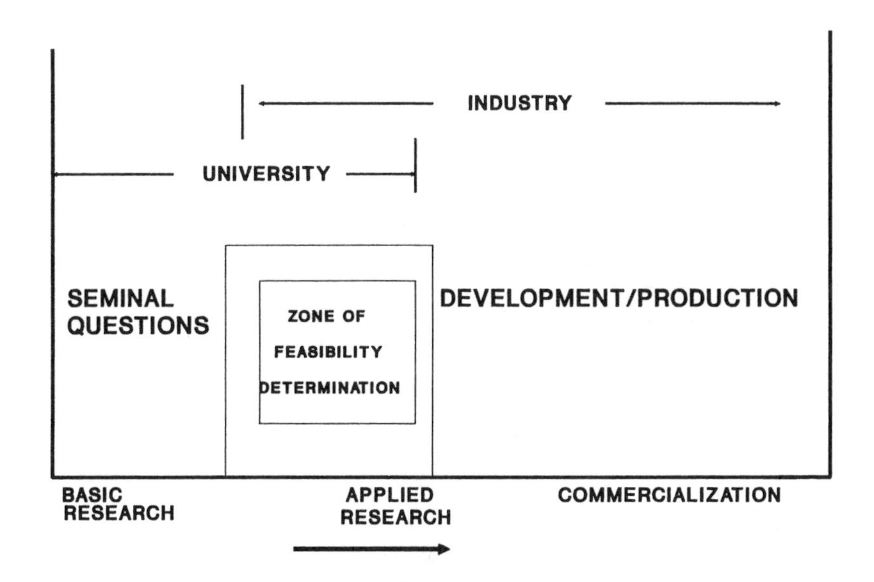

Figure 2. Innovation Process, University-Industry - Reproduced with permission from ref. 14. Copyright 1980 National Commission on Research.

on the right materials, or the processes were studied under unrealistic conditions, or the process modeling being done had neglected some important variables."

In EPIC, the research program is established through a process which is intended to increase the chances for smooth technology transfer. The companies indicate general areas of interest at the outset, and the faculty members are encouraged to be creative and to submit projects within these areas based on their experience and interests. Any proposal must consider not only scientific merit, but also the potential commercial application of the research. While this is a difficult consideration for some academic researchers, the EPIC charter states that only projects with commercial potential as well as scientific merit will be funded. A Project Proposal Review Committee makes project selections from proposals submitted by faculty members and this Committee has only member company representatives--no academics--to ensure that the commercial potential requirement is addressed.

EPIC functions with a small central staff, responsible for: Technology transfer, including the research project selection process; marketing and membership; small company relations and education and training. This staff reports to a large (23 members) Board of Trustees which is balanced among representatives of member companies, university faculty members and leaders from Northeast Ohio community organizations.

Funding for EPIC comes from the state and from membership fees. The state funds are used mainly for administration and non-research activities, as well as to support research, while the membership fees are used exclusively to fund research. Since EPIC's beginning in 1985, the state investment has been more than $13 million and membership fees have totaled about $5 million. In the early years, most of EPIC's funding came from the state, but company support has increased as more firms have become members, and membership fees now account for about one-half of EPIC's funds. The balance is expected to continue to shift toward greater private sector funding.

EPIC membership now includes more than seventy companies, about evenly divided between large, multi-national corporations who are interested primarily in the research program, and smaller regional companies who call upon EPIC to help them in solving problems, usually of a short range nature.

The research program is divided between generic projects and directed or proprietary research activities. The results from the generic research program belong to all member companies, while directed activities are for the benefit of individual companies. For example, directed funds may be used to contract for research directly with an individual faculty member, or for fellowships, or for work performed in special EPIC supported applied research facilities, which are located at the universities and are financed by EPIC.

The membership fee schedule for research investing members is based on total company sales, with an upper limit of approximately $50 thousand per year (three year commitment) for companies with sales exceeding $1 billion. A separate fee schedule, at significantly reduced rates, is used for the small member companies, who are generally not interested in investing in research.

The 50:50 formula for dividing the membership fee of research investing companies between generic projects and directed activities has been very successful for EPIC. Companies who have prior relationships with one or both of the universities can use part of their fee to continue directed activities, while obtaining significant leverage on the generic portion of their investment.

The generic research portfolio presently includes over sixty projects, representing an investment of $5 million, and it is increasing by $1 million per year. Thus, it represents a significant bank of polymer science and technology which members can draw upon. Eventually most of the research is published, but member companies get an early look at the research results, and EPIC and the universities have an agreement that patent rights will be protected before publication is permitted.

Many of the small member companies find EPIC's special facilities at the universities to be particularly helpful. These were established with EPIC support and include: (a) The Applied Research Laboratory, offering testing and analytical services, and now serving more than 300 clients per year; (b) A Blending and Compounding Center, providing a wide selection of production scale blending, compounding and polymer processing equipment, available to companies who want to study the performance of materials in different types of equipment; (c) A Mini Pilot Polymerization Plant for preparing larger than bench scale quantities of new materials for advanced testing and development; and (d) A Macromolecular Modeling Center, available both to university and industry researchers, which provides computers and software programs for mathematical modeling studies of new polymers.

Intellectual Property: Who Owns What?

The management of intellectual property resulting from generic research presents a major challenge because of the different interests of the participants. The universities have offices charged with obtaining revenues through licensing of patents. EPIC is interested in recovering some portion of the investment made by the state. Finally, the member companies, all of whom are entitled to negotiate for rights, and each with an objective of establishing the strongest possible competitive position, want patent licenses on the most favorable terms possible. EPIC has tried to accommodate the different interests of it's academic and industrial partners by a shared benefit approach. EPIC owns all patent rights, but shares any royalties equally with the university, and all members have rights to negotiate for the patented technology.

A detailed Patent and Licensing Policy describes the various options a Member has in negotiating rights. This includes funding additional research at the university, sometimes as part of a small group of companies, as well as taking the project in-house. In all cases, an agreement on license rights is required and EPIC handles the negotiation of the agreements with companies. Member companies are given the first opportunity to license a patent; but, if no member company expresses interest, other companies will be solicited.

Regarding the standard corporate objective of establishing an advantageous competitive position, any consortium faces a dilemma: Although collaborative research has its advantages, when all member companies have rights to any technology which shows promise, a company often feels it has given up the opportunity to develop a proprietary position.

The EPIC approach to this dilemma is to form a "mini-consortium" at the time a project has reached a point where technology transfer is to be considered. For most projects, this is after two or three years of university research. At this point, if a project shows promise, it is time to determine whether or not a company, or group of companies, has sufficient interest to sponsor the additional funding needed for advanced development and commercialization. Members are polled regarding their interest in the project. Those expressing interest then work with EPIC, the patent attorney and the university to develop an agreement under which the additional funding for technical work is to be provided, and licensing arrangements concerning the EPIC patents can be outlined. Alternatively, a company may elect to continue the research within its own R&D laboratories, with the objective of building a competitive position independently. A key point: companies who do not indicate interest in continuing work on a project, either as part of a mini-consortium or independently, waive all rights to any technology and patent rights which may be available on the project at the time.

The mini-consortium approach effectively narrows the field of participants from the entire membership to only a few companies, thereby offering the promise to a participating company of being able to establish an advantageous competitive position.

The results to date have been encouraging. It has been said that it is easier to turn dollars into good scientific results than it is to turn good scientific results into dollars. The EPIC research program has been at a significant level for only a few years and, considering that the time required for moving from the initial experimental work, through development, to a final commercial result is usually several years in the polymer field, it is apparent that the research pipeline is beginning to yield promising results. To date, eight patents have issued, two licenses have been signed, and several mini-consortia are under consideration.

Requirements For Success

There are three keys to success in the EPIC approach. One is the upfront requirement that there be endorsement from member companies of a proposed university project before it begins. At first this was viewed by some faculty members as an intrusion of their rights to select their own research topics--a co-opting of academic freedom. However, as the relationship evolved, it became clear that it did not follow that a research project offering the promise of eventual commercial application necessarily required a lower order of research inquiry.

As David Osborne points out (*7*), "Without links to the marketplace--without a process by which research advances are transferred into new and improved products and processes by local corporations--a strong research base adds very little to the economy."

Another key requirement is personal interactions. In the early years of EPIC, a common criticism from industrial representatives was that EPIC simply entertained proposals as submitted by faculty members, and then tried to select those which offered "the greatest promise," but without much real regard for commercial markets. However, by creating direct contact opportunities, through visits of company researchers to university laboratories and vice versa, through the Proposal Review Committee, Advisory Committee activities (about 35 corporate representatives participate on various EPIC committees) and the Semi-Annual Technical Review Conferences, bridges have been built which increase the relevance of projects to industry interests. As Mary Good stated in a recent article (*8*), "Technology transfer is as much a process of developing contacts, conduits and advocates for what you do as it is the research."

Thus, the building of bridges in the Edison Program, and in EPIC in particular, is essential in order to raise the consciousness of the university scientists regarding the possibilities which exist in fields of commercial interest for conducting cutting edge research and obtaining publishable results. In addition, financial incentives in the form of grant support in the near term and potential revenues in the future from royalties, if there is a major commercial success, have to be clearly presented.

There is complete agreement that the best method of technology transfer is through direct one-on-one communication between university and industry researchers (*9*). Written reports are very important, but they can't substitute for direct personal interaction. This personal attention not only helps bridge building for technology transfer reasons, but also increases the prospect of identifying surprise nuggets of exploitable research, something beyond the specific projects of known interest.

Finally, the "pipeline" from basic research to commercial application is always longer than we would like it to be, and it is essential that a long-term commitment be made to such programs as the Thomas Edison Program. As a corollary, it is also important to avoid raising expectations too high or too soon, and elected officials and their constituents need to be reminded periodically that these are long-term programs.

Movement by the states toward technology as a basis for economic development represents a major science and technology policy shift in the U.S. States have always had the goal of creating and attracting jobs for the benefit of their economies; but only recently, by investing in R&D, have they committed to a longer term, riskier strategy.

As former Governor Celeste stated, "The application of scientific knowledge is the basis for economic expansion and diversification, and the key to the formation of new businesses and the competitive survival of old ones." In the polymer field, Ohio has a combination of old, represented by the rubber and commodity plastics industries, and new, as advanced materials come along.

Fortunately, Ohio officials recognize the long term character of the Thomas Edison program and continue to provide substantial financial support to the Edison Centers.

Overall Evaluation: The Edison Program and EPIC

While many of the state programs are young and it is too early to measure their success in a precise way, Tornatzky and others argue that, "it is time to begin learning from the experience of the past several years in a systematic way, to try to sort out what distinguishes successful programs from failed programs (*10*)."

The report from the National Research Council (*5*) states, "The committee finds the Edison Technology Centers Program to be generally healthy, vigorous and well managed. The program is valued by Ohio's industrial community and is making a significant and growing contribution to industrial competitiveness in the state."

One of the successes of the Edison Technology Centers has been the active participation of more than 600 corporations, despite the fact that membership fees in the centers range up to $60 thousand per year, some of them with three year commitments. This says that the program meets the key market test: It promises and delivers a product for which the private sector is willing to pay.

In EPIC, it is apparent that the major polymer companies are impressed with the quality of the program. Research investing membership has increased from the 12 founding companies to over 70 members, and the revenues from membership fees have increased more than 50% in each of the last two years.

The leveraging value of the Edison Centers was recognized in the 1990 National Research Council report, which stated that the funding for the centers "is very small in comparison with the overall technical research fund expenditures in Ohio, and even small in comparison with the gross industrial product it seeks to enhance. (But) the application of these funds is seen to be well leveraged and productive, albeit over a long time span."

States have few incentives to promote cooperation within industries which are national or global in scope; but when large segments of an industry are clustered within one state's borders, the performance of that cluster has a significant impact on that state's economy. This describes the "Polymer Valley" in Northeastern Ohio. Although polymers is certainly a global field, the concentration of academic and industrial activity in Northeastern Ohio makes it worth substantial investment by the State.

While corporations are beginning to ask sharper questions regarding the value of their participation in such programs, it still holds that the best research universities, such as EPIC's Case Western Reserve University and The University of Akron, represent a research resource which major companies cannot afford to overlook. The "hunter and gatherer" opportunities, as discussed by McHenry (*11*), almost demand that a window on the research activities of such universities be maintained by the large corporations.

The National Research Council report describes the activities of the Edison Program as emphasizing: "(1) commercialization, via new ventures or new companies, of research carried out at universities, medical centers and the Edison Centers; (2) transfer of new knowledge from the universities to the larger companies, and (3) providing technological resources for small-medium companies who need help in technical problem solving." But this isn't all that is required for showing success at the bottom line.

It is clear that the success of such programs as the Edison Program requires not only that the technology transfer mission be accomplished, but also that the companies receiving the technology are successful in their commercialization efforts. Richard Florida and Martin Kenney question such programs on this particular point. In The Breakthrough Illusion (*12*), Florida and Kenney state, "There is little reason to expect that American companies will be able to turn generic innovations pioneered by R&D consortia into actual products and processes, especially when they have such a difficult time doing this with their own R&D innovations."

Other recent publications have also questioned the R&D consortium approach. Irwin Feller (*13*) notes that Tornatzky and Wykoff have challenged state programs as unproven and with serving more of a political purpose than an economic one. In other words, those responsible for putting them in place can point to patents, growing membership lists, some new firms, and also recite numerous anecdotes, but the bottom line has not yet been affected significantly. This is not unlike the situation in companies where longer range programs frequently require an optimistic and patient CEO, or an effective salesman as Research Director, or both. The state programs, most of which started in the 1980's, are long range in outlook and a realistic expectation is that pay-offs will come in the 1990's.

In fact, the Ohio program has been very productive; Coburn has reported the following accomplishments of the Edison Centers: $56 million in new federal research funds obtained for Ohio; $8 million of venture capital attracted to Ohio; establishment of a biotechnology industry base; the polymer industry expanded and strengthened; location of the largest welding center in North America; a biomedical sensor center established and establishment of one of three national centers for manufacturing sciences. Also, Kent State University, in collaboration with The University of Akron and Case Western Reserve University, and assisted by EPIC, received an NSF grant in 1991 to establish a new Science and Technology Center in the field of liquid crystal optical materials.

Overall, the Thomas Edison Program is recognized as one of the most successful examples in the U.S. of state, private sector, university and community interests working together.

The Future

Feller (*13*) points out the growing restiveness between research and education. The priorities of basic research and graduate education on one hand and commercially useful technology on the other are very different and it is necessary to maintain a balance. However, because of the pressure of competing for funding, Feller states, "in their race to be coupled with the State's economic development train, universities are presenting themselves as engines of economic growth. In doing so, they run the risk of side-tracking themselves".

One challenge facing the Edison Program Centers, including EPIC, is to strike a reasonable balance between providing too much definition of commercial needs and opportunities and not enough to persuade a faculty member to veer away from his or her personal research interests. Providing too much definition, while tempting as a means of enhancing the technology transfer process, runs the risk of unduly restricting the faculty member, or causing frustration if the commercial target is appealing, but does not offer scientific challenge sufficient to assign to a graduate student. The primary mission of the university, after all, is education.

The growing influence of science and technology centers will impact the education process at the graduate level in other ways. For example, it has been pointed out that faculty members who participate in a collective effort, as represented by a center, will have a more difficult time demonstrating that they are qualified for tenure. While a center proposal may represent a collection of research projects by individual faculty members, the director may be perceived as the only source of ideas.

State programs, because they are tailored to meet needs on a regional basis, have become recognized as a very important complement to the federal programs. For example, NSF recently announced a "State/Industry University Cooperative Research Centers" program, in cooperation with the National Governors Association Science and Technology Council, which is, in part, modeled on the Edison Program. Support for state initiatives from the federal government is expected to increase in the future.

Conclusion

The full benefits of the Edison Program are just beginning to be realized, and this is probably true of other state programs as well, most of which have been started within the last five years. This commitment by the states to technology as a means to foster economic development is a significant science policy development of the 1980's, and while the state programs engender considerable optimism, much of the return-on-investment will not be realized until well into the 1990's.

Literature Cited

1. Innovation and U.S. Research, ACS Symposium Series, The Economics of Innovation, Chapter 7 (1980).

2. Technology Administration, U.S. Department of Commerce, Room 4418, 14th and Pennsylvania Avenue, N.W., Washington, D.C. 20230.

3. State Technology Programs in the United States, Governor's Office of Science and Technology, Minnesota Department of Trade and Economic Development, July, 1988.

4. Beyond The Rhetoric: Evaluating Industry-University Cooperation In Research and Technology Exchange, Washington, D.C., Business-Higher Education Forum, 1988, Vol. 1.

5. Ohio Thomas Edison Centers: A 1990 Review, National Academy Press, 1990.

6. Proceedings from a conference at Purdue University, Digital Equipment Corporation Technology Strategy Group, 1988, pg. 52.

7. State Technology Programs: A Preliminary Analysis of Lessons Learned, The Council of State Policy and Planning Agencies, Hall of the States, Washington, D.C., November, 1989.

8. Fortune Magazine, Turning R&D Into Real Products, July 2, 1990, pg. 73.

9. Productivity Improvement Research Section, NSF Division of Industrial Science and Technological Innovation, The Process of Technological Innovation: Reviewing the Literature, Washington, D.C., Business-Higher Education Forum, 1988, Vol. 1 and reference 5.

10. Lewis Tornatzky, Proceedings of Lessons Learned: A Conference on the Strategy and Tactics of R&D-Based Economic Development, Ann Arbor, June, 1988.

11. K. W. McHenry, Five Myths of Industry-University Cooperative Research--and the Realities, Research•Technology Management, Vol. 33, No. 3, May-June, 1990, pp. 40-42.

12. Richard Florida and Martin Kenney, The Breakthrough Illusion, Basic Books, 1990, pg. 180.

13. Feller, What's Happening at the State Level, presentation for the National Conference on the Advancement of Research, Chicago, September, 1990.

14. National Commission on Research, Industry and the Universities: Developing Cooperative Research Relationships in the National Interest, 1980, pg. 5, Figure 1.

RECEIVED April 5,1991

Chapter 5

Policy and Collaboration in Research and Education

New Alliances

Don I. Phillips

Government–University–Industry Research Roundtable, 2101 Constitution Avenue, NW, Washington, DC 20418

In this paper, I will address two different perspectives on government-university-industry relationships. The first is collaboration on matters of policy through the operations of the Government-University-Industry Research Roundtable. I will describe the purposes, essential features, and accomplishments of the Roundtable. The second is programmatic collaboration in research and education. Here I will present the principal observations on university-industry cooperative programs that have emerged from several Roundtable activities on this topic.[1]

The Government-University-Industry Research Roundtable, sponsored by the National Academy of Sciences, National Academy of Engineering, and Institute of Medicine, was established in 1984 as an alternative to the traditional study commission approach to science and technology policy issues. The vision of all those involved in the initial concept was that the Roundtable "could accomplish what present structures have been unable to do" in addressing major policy issues, problems, and opportunities in the research associations between universities and their principal partners, industry and government.

---

[1]The views expressed here are my own, and not those of the Research Roundtable nor its sponsoring organizations, the National Academy of Sciences, National Academy of Engineering, Institute of Medicine.

0097–6156/92/0478–0042$06.00/0

Today, seven years later, the Research Roundtable has evolved from the initial vision through a stage as an experimental, "entrepreneurial" venture, to its present status as an established, unique organization in national science and technology policy affairs. The essence of the Roundtable, and its uniqueness, is that it is the only forum in the United States where the leaders from all sectors of the research enterprise talk to one another on a continuing basis and in a structured way about promising opportunities and issues that challenge, trouble, and, occasionally, divide them. Maintaining the original intent, the emphasis is on understanding issues, examining all the relevant perspectives on a topic, and seeking common ground among all the parties. The Roundtable structures and illuminates issues, it does not decide them; nor does it make recommendations or offer specific advice.

The focus of the Roundtable is on the major institutional, organizational, and policy issues affecting American science and engineering. The stewardship of the research enterprise is the Roundtable's brief, and it was organized on the premise, innovative in 1984, that all sectors—federal and state governments, universities, and industry—share the responsibility for that stewardship.

The make-up of the Roundtable Council, which guides the overall operations, illustrates the point . Included are the senior federal R&D officials, a current and former governor, senior officials from academia and industry, and working academic scientists and engineers. It is the participation of the senior federal officials as full sitting members of the Council that distinguishes this group from others that address similar issues. The Council sets the agenda for the Roundtable and selects the topics to be addressed.

How does the Roundtable work toward accomplishing its objective of promoting mutual understanding and dialogue among the leaders of the American research enterprise? The action of the Roundtable is based not on reports with specific recommendations and advice to government officials, but rather on the Roundtable's ability to get the right people in the right room at the right time, supported by appropriate background and analytical information, to inject new ideas and deeper understandings into deliberations on research policies and procedures. The starting point is the Council meetings themselves. Because the Council participants hold leadership positions in a broad range of organizations, the intent is for the insights gained within the Council to be translated into the operations and policies of the government, academic and industrial sectors. The strategy also includes convening working groups and devising outreach activities to stimulate discourse in the broader research community. The centrality of the Council and working group meetings to the Roundtable's objective was best stated by a federal agency head who is a member of the Council: "Don't judge the

Roundtable just by its products; its most important contribution is the exchange of ideas and perspectives that takes place during our meetings."

What has the Roundtable accomplished over the past six years? The emphasis on dialogue, supported by background and analytical information, has produced a remarkable track record. One way to summarize the Roundtable's accomplishments is with the topics that have been addressed.

Remember, however, that these Roundtable reports are not ends in themselves but are a means for bringing ideas to the groups and individuals who can take and shape actions. In working toward this objective, the Roundtable has:

- increased the involvement of new communities, notably industry and state officials, in science and technology policy deliberations; examined university-industry alliances and established a Federal-State Dialogue on Science and Technology;

- convened frank discussions, when tensions were especially high, between university administrators and faculty and senior officials of the Office of Management and Budget on controversial issues in research funding and management;

- stimulated the design, testing, and implementation of new, streamlined procedures for the management of federal research grants to universities through the Florida and Federal Demonstration Projects;

- provided a forum for the analysis and discussion of issues before they arose elsewhere—especially in the areas of science and engineering talent and the status of the academic research enterprise;

- convened discussions on the impact of federal budget constraints on the research system;

- increased the understanding of research facility financing and promoted discussion of cooperative approaches among federal, state, and university officials.

What is responsible for the Roundtable's success to date? And, what is required to maintain its effectiveness in the future? One answer to these questions stands out from all the others—that is, the full and active participation by the senior federal R&D officials. They contributed at the outset to the establishment of the Roundtable; they have taken on leadership roles in the Council and the working groups; and they are sharing responsibility now with other Council members for deciding how best to maintain the vitality of the Roundtable and the relevance of its agenda. Most importantly, this commitment has spanned several changes in the participating federal officials. As new persons have been appointed to the senior federal R&D positions, there has been an easy transition to their involvement in the Council and other Roundtable activities.

While the participation of the senior federal R&D officials is at the core of the Roundtable's past and future success, other features of the Roundtable structure and operations also have been and will continue to be important. They include:

- Neutral Setting. The sponsorship by the National Academy of Sciences, the National Academy of Engineering and the Institute of Medicine provides a neutral setting with credibility among all elements of the research community in the three sectors.

- Continuity vs. New Initiatives. One of the unique features of the Roundtable is its capacity for continuity and follow-up on issues. While maintaining this characteristic, the vitality and effectiveness of the Roundtable also depend on its undertaking new initiatives. The Roundtable achieves a balance between follow-up activities on current topics and new projects.

- Long Term vs. Short Term Issues. The Roundtable maintains a balance between attention to broad, ongoing concerns of the research community and to a search for solutions to immediate problems.

- Addressing Problems from both Policy and Operational Levels. The combination of study and analysis by operational level representatives in working groups and discussion by policy level representatives in the Council has produced an environment that leads to the introduction of new ideas and new procedures into the research system.

- Balanced Views. All points of view are presented in Roundtable deliberations. The Roundtable has avoided becoming a proponent for the views of any one constituency.

- Flexible Financial Support. Support for the Roundtable has been provided by foundations, federal agencies, industry, universities and state agencies. The majority of these funds is provided as general support for the Roundtable, enabling the Roundtable to respond quickly to problems and opportunities as they arise and to address issues in flexible, diverse, and innovative ways.

- Personalities. The Roundtable is foremost a process—a process for bringing together the diverse constituencies concerned with the research enterprise. The ability of the Roundtable to stimulate constructive change in the system depends on the "delicacy" and the balance with which it is able to address issues that are typically complex, intractable, and controversial. As such it is an intensely personal enterprise, whose effectiveness has depended on the ability of the Roundtable Chairman, the Council, the Working Groups, and the staff to work constructively with the full range of relevant constituency groups and individuals.

The evidence of the past seven years demonstrates that the Roundtable does contribute in unique and valuable ways to "improved communications on important issues of policy" and to more effective working relationships among the sectors. Extraordinarily busy and talented people from all sectors have committed themselves to the work of the Roundtable, believing that it offers the single best instrument for addressing the stresses on the research system and for maintaining the vitality of our science and engineering enterprise. Furthermore, assuming there is credence to the adage, "imitation is the highest form of flattery", it is noteworthy that the Roundtable has served as a model for the start up of similar organizations within the Academy Complex, elsewhere in the United States, and in other countries.

Perspectives On Industry-University Collaborative Programs

The Roundtable has been examining and promoting discussions on industry-university alliances since it was established. The initial project mapped the diversity of research alliances, reviewed the principal issues of controversy and debate, and culminated in a national conference and report.[2] In 1988, the Roundtable together with the Industrial Research Institute published model agreements for university-industry cooperative research to serve as starting points for negotiations between industry sponsors and universities on

---

[2]New Alliances and Partnerships in American Science and Engineering, Government-University-Industry Research Roundtable, 1986.

grants and contracts.[3] Recently, we carried out an assessment of the usefulness of these models.[4] And, we just completed interviews with seventeen senior industrial research managers to obtain their perspectives on innovation and on how alliances with universities are expected to contribute to technical change and competitiveness within individual companies.[5]

The observations on industry-university alliances that follow are based on these activities and reports. Three themes stand out:

- Industry-university cooperative programs are a continuing series of experiments characterized by a great deal of variety and diversity in form, content, and objectives.

- Most of the cooperative programs are able to work out acceptable arrangements for dealing with financing, publication, communication, patent ownership, faculty roles, and many of the other features that caused much initial controversy. Intellectual property and licensing agreements continue to be a source of much difficulty, however.

- Based on our interviews with senior industry officials, I conclude, somewhat surprisingly, that there is still a gap in understanding between universities and industry on the purposes and expectations of industry-university alliances.

Dimensions of Industry-University Collaboration

All is Not New. Commentators sometimes write as if relationships between universities and industry were totally new. In fact, recognizable antecedents go far back in time. For example, academic chemistry has, from

---

[3] Simplified and Standardized Model Agreements for University-Industry Cooperative Research, Government-University-Industry Research Roundtable, 1988.

[4] "Survey to Assess the Usefulness of Two Model Agreements for University-Industry Cooperative Research", Government-University-Industry Research Roundtable, 1990.

[5] Industrial Perspectives on Innovation and Interactions with Universities: Summary of Interviews with Senior Industrial Officials, Government-University-Industry Research Roundtable, (forthcoming, Spring 1991).

the beginning, been closely tied to industrial chemistry. Much of modern biology is also deeply rooted in the search for problems. Similarly, computer science is closely tied to applications. And, of course, the set of applied scientific fields which call themselves "engineering disciplines" are also by their origin and their nature oriented to applications. Propositions about a natural chasm between academic science and industrial science have often been drawn too sharply and too globally. Indeed, academic science and industrial science in the United States grew up together.

Variation and Diversity. Diversity in companies and in universities results in great variety in the nature, type, and objectives of industry-university interactions. University cultures vary as do their attitudes towards the kinds of relationships with industry that are or are not appropriate. Those institutions with long standing liberal arts traditions tend to avoid relationships other than those that support basic research. The technical universities have shown a greater willingness to engage in applied research with industry funding, a greater respect for the proprietary interests of the funder, and a greater interest in continuing close interactions with industry. Companies also differ in their views toward research, toward in-house and externally sponsored research, and toward collaboration with other companies and with universities.

Given this cultural variation, it is not surprising that the new partnerships vary considerably in the kinds of activities and arrangements that are involved. Some are largely concerned with basic research. In other arrangements, the purpose of the work is to solve a well-defined practical problem. Training of undergraduate and graduate students may or may not be part of the program. Consulting by the involved university personnel is in some cases restricted, but in others, consultation is an important aspect of the arrangement. Similarly, in some cases constraints are imposed to limit faculty entrepreneurship, while in others the arrangement is designed to channel or facilitate entrepreneurship.

Financial Support. Overall, corporate support for university research will perhaps never exceed 7 to 8%. Industry funding for university research comes largely from corporate research budgets, which are nearly always quite small relative to development budgets and are likely to remain so. Still, corporate funding is significant at some schools, reaching levels of over 20%, and is more prominent in some fields than in others, notably, semiconductors and biotechnology. There is concern about the sustainability and the breadth of industrial funding. The new alliances are concentrated in a few industries, for example, biotechnology, microelectronics, and special materials. Will sufficient short-term results materialize to maintain industry's involvement with universities over the long-term, even as the fields of interest may change? We currently see signs of changes in industrial support for research and development (R&D), both in-house and externally. Industrial support for academic R & D must be considered as a complement to, not a substitute for, federal support. The general view is that federal funding of academic research is critical, both for the long-term vitality of research and graduate education and for attracting industrial support.

Industry-University Symmetry. The capacity of a company to assimilate advances in research is related to the internal technical capabilities of the company. A breakdown in symmetry between the technical capabilities of cooperating companies and universities will inhibit the ability of the company to transfer innovative ideas into technology. Internal industry R & D is an important component of technological innovation, and industry must maintain its investments in in-house research if it is to benefit from participation in collaborative programs with universities. Participation in such programs by industry cannot be viewed as a substitute for internal industry R & D.

University Concerns

Strategic Role for Universities. The alliances are making the lives of universities more complicated and more exciting. As a part of these new alliances, universities are assuming visible and explicit strategic roles in state, federal and industrial economic and technological development programs. This has resulted in increased expectations being placed on universities and in greater political currency given to university affairs—developments that have produced both strains and benefits within the university community. Strains are caused by differing views of new university activities tied to industry and by the increasing political interests in universities as indicated by special appropriations by the U.S. Congress for university research facilities and programs. Benefits come in the form of new state and industrial investments in university programs and the excitement resulting from the opportunity to work with new people and on new scientific and technical problems. Reaching the right balance in these forces on the universities will require care, nurturing, and thoughtfulness by the universities themselves and by the patrons and policy-makers that influence universities.

Industrial Influence on Academic Research. A major concern raised by university-industry cooperation is that corporate values will divert academic research from its proper role, the search for knowledge. It does not appear that this is occurring. University and industrial participants are in the main agreeing on the research that warrants support. One view is that a major cultural change in universities came after World War II, when agencies like DOD and NIH began to support "really fairly directed basic research. " In this light, industrial support is only "a small perturbation."

Faculty Loyalties and Incentives. There has been a change in faculty loyalties over the past forty years. Prior to World War II, little funding was available outside the university, and faculty concerns were directed toward their own institutions. With the significant increase in federal support, there came incentives for promoting individual disciplines and growth in professional and scientific societies. Faculty loyalties were directed toward their disciplines, their colleagues in the relevant societies, and their program officers in the federal funding agencies. Now, the potential for significant increase in academic salaries through alliances with business and the financial community may diminish faculty loyalties to their universities and their disciplines. To some this is a major concern; others see this as the exception rather than the rule. They see faculty loyalties to science and engineering running high in spite of the possibility for individual financial gain.

Freedom of Communication. The alliances do not appear to be imposing unacceptable constraints on publication and communication, except perhaps in highly competitive fields like biotechnology. Here, however, views differ as to whether these constraints are brought on by commercial or scientific competition. In one sense, industrial-academic connections have served to increase communications among scientists and engineers between sectors and between disciplines.

Industry Perspectives[6]

Innovation and Technical Change. Innovation is considered as the movement of an idea from its conception to a commercial success, either as a product or a process. Technical change and technical advance are steps in and contributors to the innovation process. In many but not all industries, most innovation occurs through in-house incremental improvement to existing products or processes rather than the rarer breakthrough event that revolutionizes a product or process. Industrial approaches to technical change and collaboration with universities are based on this perspective. It is the universities that are at the forefront of scientific innovation, but it is within companies that product- and process-oriented technical change occurs for most fields. The limited role of universities in innovation has not been recognized because of the misconception that technological change generally

---

[6]The views represented here are those of the seventeen industrial officials recently interviewed by the Roundtable; op. cit.

occurs through a remarkable breakthrough that will revolutionize an industry; because of the excitement that accompanies such radical new ideas regardless of how infrequently they occur; and because university scientists tend to have a simplistic understanding of how product development and commercialization occur.

<u>University Roles in Innovation</u>. The primary role for universities in innovation is as educator and provider of talent. Included here are the essential roles universities play in cooperative and continuing education. Providing new knowledge and in-depth understanding for scientifically and technologically new or emerging ideas also are significant roles for universities. Industry-university collaborative programs should be established with a clear understanding by both parties of these education and research roles for universities. Universities should not, in the view of the industry officials, attempt to move their research into arenas closer to product discovery; this is not an appropriate role for universities, nor is it a task for which they are generally well-suited. Beyond education and research, the process of innovation also depends on a complex network of interactions and exchange of ideas, and universities are a central part of the glue that holds the network together. If the university research system did not exist, information would not flow well.

<u>Generic Research</u>. Most of industry officials expressed skepticism of the results of generic research for competitive advantage and thus are not willing to support it to a large extent; they believe that such support is an appropriate role for the federal government. In the interest of national welfare, many firms are willing to spend a small amount—less than 2% of their R & D budget—on this type of research, although there is some variation by industry. In addition, precompetitive, generic research is not generally pursued through company financed consortial arrangements with universities either. Companies would want to retain proprietary status for any discoveries that do emerge, and would prefer not to lose competitive advantage through the sharing required in the collaborative programs. Companies will collaborate with each other in some areas of precompetitive research, however, as a way to leverage expenditures and to advance technologies to new levels of understanding.

<u>Collaboration with Universities</u>. Although most firms look to themselves to be the source of incremental technological advance, all interviewees acknowledged their reliance on universities, as well as on other organizations, for scientific and technological breadth and in-depth understanding. Industry needs new knowledge from universities in order to build new technologies and to improve upon old ones. In general, industry interviewees stated that they find the most fruitful form of collaboration to be a bottom up approach based on one-on-one relationships between

industry and university scientists. Arrangements with consultants and informal interactions also are strongly supported. Industrial officials were less enthusiastic about collaborative arrangements through university centers, consortia, or affiliate programs. They see these as mechanisms for accessing expertise—consultants or recruits, for promoting good will, or in some cases for promoting basic science or addressing generic issues in technologies, but not as arrangements closely linked to commercialization: they are too remote from the market place, they have uncertain benefits, and there is a strong motivation for individual industrial members to develop products in-house rather than share. Likewise, the industrial officials were skeptical about the value of large multi-million dollar partnerships with universities or university departments. Some commented that while the projects are viewed as great successes from the university perspective, from a profitable perspective, their success remains to be seen.

Intellectual Property Rights. As would be expected from the industry perspectives summarized above, industrial officials have strong views about the roles of intellectual property rights and patenting and licensing agreements in university-industry partnerships. In most industries, the probability of any commercially viable intellectual property evolving from an alliance with a university is remote, they believe, because of the type of research that companies tend to support at universities. At the same time they observe universities becoming more stringent about their intellectual property rights and expectations of financial gain. This behavior is causing divisiveness on the campus and between the university and industry participants, according to interviewees. In our survey of users of the model agreements, university officials also identified intellectual property rights as a contentious topic in industry-university relations.[7]

In-House Collaboration. Many of the industry interviewees noted that the problem of bridging between research ideas and product development that occurs in industry-university collaborative research is also present within a company; the same break downs in communication and understanding can occur. Some interviewees stated that the research divisions of their firms even view themselves as part of the academic system.

Expectations of Industry-University Alliances

What should we expect of industry-university alliances? In addition to the above observations, perspectives of Ralph Gomory, formerly senior vice president for science and technology at IBM, and Harold Shapiro, president of Princeton University, are relevant in answering this question.[8]

---

[7]"Survey to Assess the Usefulness of Two Model Agreements for University-Industry Cooperative Research", Government-University-Industry Research Roundtable, 1990.

According to Harold Shapiro:

> . . .if we want to get economic growth out of new science and technology, we have to pay attention to what I call "everything else," and the everything else really could not be summarized better than by saying "how groups work together"—how we relate to each other how we treat each other, and how we trust each other.
>
> . . .what may not be in our best interest is the belief that superiority in science alone—at the expense of "everything else"—will ensure this country's economic strength.
>
> The lessons of history tell us otherwise. For example, it was not Britain's science and technology superiority that made it first in the Industrial Revolution. It was political stability, it was the society's concept of private property, it was decentralization of authority in British institutions. It was not that the British had better science than in Belgium and France. It is very, very seldom that a monopoly on science alone has produced a tremendous spurt in sustained economic and social dividends. Why is it that we do not read that lesson?

According to Ralph Gomory:

> . . ."pull" [by a company] consists of people who know what they need going out and looking for it—and finding it—in a vast universe, rather than asking outsiders who don't know the company's situation to throw pieces at it. "Pull" is much more likely to succeed, moreover, because the burden of finding uses for research belongs not with universities but with the companies themselves.

---

[8]Gomory and Shapiro, **A Dialogue on Competitiveness**, ISSUES IN SCIENCE AND TECHNOLOGY, SUMMER 1988 (National Academy Press); GOMORY AND SCHMITT, **Science and Product**, SCIENCE, May 27, 1988.

> A strong science base . . . cannot make up for inadequacies in the functioning of the development and manufacturing cycle [within companies].

The results of the Roundtable activities together with these views lead me to think of industry-university alliances in the following manner. The industry-university alliances should be viewed as a new and creative way to contribute to excellence in both academe and industry and not as the major national effort to solve our competitiveness problems. The nature of research, of technology development, and of education is changing in many areas of science and engineering—particularly those areas, for example, electronics, biotechnology, and materials, around which many of the alliances are forming. These changes reflect the fact that boundaries between the underlying disciplines and between basic research and applied research are blurring, advances in fundamental knowledge are becoming relevant to technology development in the near term, R & D are dependent on and in some cases limited by sophisticated and expensive instrumentation, talented scientists and engineers are in short supply, and product life cycles are becoming shorter. Within this environment, maintaining research capacity at the frontiers of knowledge and maintaining technological capacity at the frontiers of product and process innovation require greater collaboration and interaction between academic and industrial scientists and engineers than has been the norm.

The emerging new alliances, therefore, are essential to maintaining the nation's scientific, technological and educational base. To the extent that this base contributes to our international economic competitiveness, the alliances are an important part of the strategy. But, we know that the strategy for economic competitiveness must include many other factors of equal and perhaps even greater importance.

RECEIVED May 6, 1991

Chapter 6

The Du Pont Honors Workshop

A Successful Industry–School Partnership

John W. Collette, Sharon K. Hake, and Robert D. Lipscomb

E. I. du Pont de Nemours and Company, Wilmington, DE 19898

Since 1984, Du Pont has held summer workshops, for high school teachers, designed to raise the quality of pre-college science teaching and foster industry/school partnerships in communities near Du Pont operating facilities. The workshops provide an introduction to polymer technology and biotechnology through lectures, hands-on laboratory experience, and field trips. An active follow-up program with graduates of the course has been established to build a continuing relationship with the technical community. Evaluations show that the workshop experience provides: enhanced professionalism, a better understanding of industrial chemical research, and more pertinence to their teaching.

The Du Pont Honors Workshop for high school science teachers is one component of Du Pont's corporate support for education, a program which dates back more than 70 years.

Historically, this program has contributed money primarily to those colleges and universities which are important in our hiring. It is given in the form of unrestricted grants, young faculty grants, and support for minority education. In the last decade, however, we have added support both for pre-college education and for programs aimed at improving the public understanding of science and technology, such as for exhibits at the Smithsonian and at the Franklin Institute in Philadelphia. In 1989, total contributions in these programs were around $19MM, independent of the many direct research collaborations in which our research staff participates.

Pre-College Program

In the early 1980's, a series of national reports issued *(1)*, culminating with the publication of "A Nation at Risk", which focused attention on the serious

0097–6156/92/0478–0055$06.00/0

problems of pre-college science and mathematics education *(2)*. These reports identified the need for industry to become involved in improving the situation.

After a thorough study and consultation with local teachers groups, Du Pont initiated a program aimed at improving pre-college science teaching. This began as a small effort in 1984 and had grown to almost $3MM by 1989. About 70% of this is given corporately through the programs discussed below. The remainder represents contributions by more than 90 individual sites to their local schools. In Boston, Du Pont is one of ten industrial sponsors of PROJECT 96, an effort in which Boston University and downtown schools are working with 100 children due to graduate in 1996.

The pre-college effort initially focused on secondary and high school science, with the objective of attracting more students to science and engineering careers. As we gained more experience, we, along with many others in the country, have come to recognize that high school is too late to begin to interest students in science--and especially in science careers. By that time the students may have been turned off by poor teaching, or they may lack some of the fundamentals needed to proceed on a technical course of study due to lack of exposure to science education in earlier years. In addition, we recognized that there was an equally important need to raise the level of understanding of science and technology among the 85% of school students who will not go on to technical careers.

As a result, our emphasis on improving elementary science education has steadily increased to where it now represents 1/3 to 1/2 of the pre-college effort.

Program Strategy

There are numerous ways to help in the educational arena, virtually all of them worthy. We have chosen to concentrate on building and working through local partnerships with schools and governments. We have a small permanent staff to handle the organizational and funding aspects, but most of the work depends on the interest and volunteer efforts of our employees.

The program has three interrelated strategic components.

The first involves educational reform. At the national level, we are participating actively on the National Board for Professional Teacher Certification. At the state level, we have various initiatives in localities where Du Pont has a significant presence. The most advanced of these is work done to upgrade management and leadership skills of school administrators and superintendents in the Texas school system. This is now being extended to several other states through the Department of Education's "Leadership in Educational Administration Development" (LEAD) program.

The second is aimed at providing better curriculum material. We support the PBS program, NEWTON'S APPLE, targeted at making science interesting and relevant for secondary school students. We also support (along with many other companies) the excellent Chemical Education for Public Understanding Program (CEPUP) being developed for Grades 7-9 by the Lawrence Hall of Science at the University of California (Berkeley).

The third component is teacher enhancement. Here we have many initiatives, one of which is The Honors Workshop, discussed in more detail below. Others are:

- Travel to National Science Teachers Association (NSTA) meetings. This began in 1984 when Du Pont sponsored the attendance of 32 Delaware teachers to the NSTA meeting in Boston for professional renewal. This had such a positive impact that the program has continually expanded since then. This year we sent 250 teachers to the NSTA meeting in Atlanta, April 4-8. These teachers came from school districts located near 67 Du Pont sites in 30 different states.

- The "Du Pont Challenge National Science Essay Contest", co-sponsored with General Learning Corporation in cooperation with NSTA. More than 10,000 junior and senior high school students participated in this contest in 1990. The winners in each division received $1500 and attended the NSTA convention in Atlanta along with a parent and a teacher. Cash awards and recognition were given also to the 52 runners-up.

- Science Alliances *(3)*. These local alliances, pioneered by the "Triangle Coalition for Science and Technology Education," serve to link scientists interested in working with schools to teachers who are looking for help *(4)*. A Science Alliance was formed recently in Delaware to serve both the state and nearby communities in Pennsylvania and Maryland. Chad Tolman, a Du Pont chemist, has taken a year's leave of absence to work as Chair of the Coordinating Committee and help make the Alliance self-sustaining. Major programs include an Elementary Science Olympiad, Workshops for volunteers and elementary teachers, Summer Industrial Fellowships for secondary teachers, and classroom visits and demonstrations by scientists, when requested by the teachers.

- Science Workshops are supported at many sites. In Delaware, for example, we support a program called QUEST (Quest for Excellence in Science Teaching), which holds seminars and workshops at which science teachers can improve their skills.

The Honors Workshop

The Honors Workshop is aimed at rewarding and improving master science teachers nationwide. A basic premise behind this approach is that an inspiring teacher can be the key factor in a student's decision to choose a science career.

The program began in 1984 as a cooperative effort between NSTA and the National Science Foundation to fund two-week workshops at which teachers could learn about industrial technology. Du Pont participated by offering to hold a workshop on polymers at the Du Pont Experimental Station, the Company's major research center, located near Wilmington.

Our first workshop had only seven teachers and was a very positive learning experience on both sides. The teachers were clearly stimulated by the course work, and our own employees were enthusiastic about the opportunity to teach and to interact with the group. We considered it so successful that we elected to continue the workshops on our own.

The course was a comprehensive introduction to polymer technology - synthesis, characterization, processing and fabrication, and diverse applications such as composites and membranes. Lectures were integrated with laboratories and tours of production facilities to provide practical experience.

One measure of the success of the workshop was a Teachers Guide to polymer chemistry*(5)*, subsequently published by NSTA, which used material assembled from our workshop (and similar ones on polymers held at Shell in Houston and ARCO in Pennsylvania). The text, written by the teachers, has an extensive selection of classroom experiments which the teachers designed and tested. This book, now in a revised second printing, has been used in training technicians and is distributed and used in both high schools and in college courses given for teachers.

The workshop has evolved a great deal since that first year, primarily as a result of periodic surveys of the graduates to see what they found most useful over time. These surveys and individual discussions gave us a better idea of the many barriers teachers face which limit how readily they can use and incorporate new material in their classrooms. Some of these are:

- Time: The curriculum is already very full, so it is unrealistic to think of introducing a segment devoted only to a subject such as polymers. Teachers can and do use the information to exemplify or reinforce the basic science they have to cover.

- Administration Support: A teacher who may be the only chemistry teacher in the school may have a hard time getting administration support for new initiatives.

- Samples and Supplies: Any hands-on experiments must be very simple, dependable, and if at all possible use available supplies and equipment. One of our teachers called it "hardware" chemistry - that is, based on things he could pick up at a hardware store.

- Expertise: Many teachers do not have scientists available as a resource if there are some aspects of the subject which he or she does not understand.

- Money: Funds are very limited and are budgeted well in advance. Even small amounts (e.g. $100-200), which can make a significant difference, are usually not available.

- A Bimodal Student Population: 20% are very interested and quite capable in science and 80% are less interested or able. Teachers have to teach both.

The workshop experience has many positive impacts on the teachers.

1990 Workshop Program

Our program today has responded to those findings. It is still focused on upgrading the teacher's understanding of current industrial chemical technology and at helping them make science relevant to the everyday experiences of their students, but we have modified the content and the approach.

We aim for a class size of about twenty, which allows a balance of individual attention and group interaction. Nominations are solicited primarily from school districts around our manufacturing sites and from teachers who have recently won Presidential awards. We have found it very effective to work through our plant sites because it helps them to build partnerships with the schools in their locale.

Nominees must be recommended by their school science coordinator and must submit a record of their educational plans. Attendees are chosen competitively by a steering committee, which includes a teacher who has attended the workshop and a Professor from the University of Delaware. This year the teachers will receive credit from the University after completing course requirements.

The course curriculum has been changed as is evident in the 1990 course outline. (See appendix.) Polymer technology remains at the core of the course but is covered in much less depth. The original curriculum was appropriate for the teachers who attended because of their interest in polymers, but we learned much of it was too specialized for a wider audience.

We have added an introduction to biotechnology as we found that many--sometimes a majority--of the teachers who attended taught only biology or both biology and chemistry; and we were fortunate in having an outstanding group of scientists and engineers carrying out research in modern biotechnology at the Experimental Station.

Engineering topics have proved very popular with the teachers as it helps them understand better what engineers do in industry. In the past this has generally involved polymer engineering (processing, design, and fabrication). In the past two years we have covered bioengineering to expose the teachers to typical problems engineers work on in this new technology.

Environmental topics are included where they fit in with the main theme of the workshop--plastic waste and recycle, for example, or ozone depletion. These are included because the teachers and their students are vitally interested in understanding the environmental issues better, what the trade-offs are, and what is being done about them.

We still keep a significant emphasis on the experimental aspects of chemistry. We aim for about equal time between lectures and laboratories or demonstrations. With the help of a local high school teacher who had taken the course earlier, we revised most of the experiments in the labs to provide more hands-on experiments which could be readily used in a typical classroom situation. This means using chemicals which are readily available and can be handled safely without excessive caution, avoiding unusual equipment, and designing the experiments so that they can be accomplished in a typical classroom period.

One experiment, designed to show the polymeric nature of cellulose, involved dissolving filter paper in cuprous ammonium hydroxide to form a viscose solution from which "fibers" could be spun with a hypodermic syringe. Polymer crosslinking is demonstrated by allowing a commercial epoxy to react in a soda straw for various times and temperatures; the viscosity increase is followed by allowing metal BB pellets to fall through the mixture. Kinetics of yeast growth was correlated with glucose uptake in the bioengineering laboratory.

Teacher Impact

The workshop experience has many positive impacts on the teachers. Probably the most important is a real growth in professionalism and a reinforcement in their commitment to science education. This is noted repeatedly in follow-up surveys. We have had several examples in which a participant attended the course despite feeling "burned out" and ambivalent about continuing teaching. In every case, the individual returned to teaching with renewed dedication.

This professionalism is strengthened by the networks the teachers form with other teachers at the workshop, which are subsequently nurtured through attendance at the NSTA meetings and through a Newsletter recently started by one alumna.

Workshop participants leave with a better understanding of industry and what is involved in industrial research. They learn a great deal about our attitude toward handling chemicals safely and appreciate that we, too, are concerned about the environment.

The teachers also report that they can give much better career guidance to their students. We do everything we can to encourage one-to-one contact with a range of our employees during the workshop. Any employee who is involved in teaching or in a demonstration usually joins the teachers for lunch or dinner so they can follow up the lecture and lab informally. As a result the teachers find they can speak authoritatively to their students about what chemists, engineers, and technicians really do in industrial research.

The Du Pont Honors Workshop promotes the idea that classroom activities should stress the relationship between science, technology and society--demonstrating how science applies to everyday life. The 1989 participants suggested these classroom activities.

- linking science to other careers
- making abstract ideas real
- using products that students are familiar with
- requiring students to think critically
- developing ideas and concepts based on reason and observation
- inspiring students to ask questions and to try some experiments themselves.

There is invariably much spontaneous feedback from the attendees after the workshop is over. Here are some typical responses.

"The Workshop was one of the most professionally stimulating programs that I have ever participated in. I cannot tell you how much I have used the information from that workshop in just two short months! I never felt that I was participating in a public relations program for Du Pont. To have the opportunity to talk with those men and women who are truly shaping the technology of today and the future was an extraordinary experience." Claudia Fowler, Christa McCauliffe and Presidential awardee, The University Laboratory School, Louisiana State University, Baton Rouge, LA.

"My life has been completely changed. Because of you I am having a wonderful school year. Thank you for giving me the opportunity to attend Du Pont's Honors Workshop in Wilmington. Delaware. I was seriously thinking about quitting the teaching profession but thanks to your company's workshop I believe there's hope for the future of our youth. Your company has stopped a "good teacher from quitting." Leevones G. Johnson, Blount High School, Prichard (Mobile), AL.

"I want you to know how much I appreciated the opportunity to attend the Du Pont teacher workshop this summer. In going over all the material I am somewhat amazed at how much we did, how much we learned, and how much fun it was participating in the program. My regret each year is that my students are the victims of a curriculum underdeveloped in science. I'm trying to figure out what to do about the situation and how to go about changing it. Again, many thanks to you and to Du Pont for contributing in a positive way towards improving science education." Marianne B. Anderson, Presidential awardee, Pocatello High School, Pocatello, ID.

"The workshop was designed as a reward for outstanding teachers, as well as to stimulate creative approaches to teaching 'real-life' science. The 1989 Workshop certainly accomplished these goals in the finest sense of the word. I want to formally thank the Du Pont Company, not only for this workshop, but for caring about the future of science education in this country and doing something about it." Carolyn Thomas, Governor's Award for Excellence winner, Circleville, OH.

Follow-Up

A key element in the success of the program is an effective follow-up effort.

Dr. Lipscomb, a retired employee with extensive polymer experience, acts as liaison to the teachers and keeps in contact with them directly or through attendance at NSTA meetings. The teachers are encouraged to contact him for information and/or material. This might include copies of the texts used in the course, samples of different kinds of polymers or fabricated parts, trade literature or videotapes.

During the Workshop, many detailed outlines along with copies of overhead and projection slides are distributed, so that a sizable reference volume is accumulated. The "Teachers Guide" to polymer chemistry, referred to above, and three texts on bio-related subjects, are given to the teachers during the Workshop. Subsequent to each Workshop, several supportive items are mailed to enable and encourage the teachers to better utilize the information gained. For example, after the 1990 Workshop, we sent eleven different "show and tell" samples of polymers along with descriptive material to each attendee. Also included were videotapes on biotechnology and on protecting the ozone layer and a tutorial disc, "An Introduction to Polymerization", suitable for use on Apple II and similar computers. Several of the teachers have requested and received additional copies of the books, especially the polymer text, for use in local workshops they are sponsoring. Over 100 copies of the polymer book have been distributed in the three months following the 1990 Workshop along with numerous additional polymer samples and literature for follow-up "in-service" programs.

This year, as a further follow up, we have established a mini-grant program to fund projects the teachers wish to undertake as a result of the workshop.

Looking Ahead

We are pleased with the results we have had to date with this program. We view it as a way to catalyze and strengthen partnerships between our many sites and the school districts around them, working in collaboration with the NSTA.

We have learned a number of things important to others who decide to undertake similar programs. First, we have found that the teachers are more than capable at adapting advanced technology to their own science courses. Second, good hands-on experiments are critical in teaching the experimental method and in demonstrating that chemistry is an experimental science. Third, individual personal contact with our employees is essential in building an understanding of industrial work. Fourth, an effective follow-up program is needed to reinforce and sustain their experience.

We plan to keep the course aimed at making the best teachers better. We recognize that 20 teachers a year seems rather minuscule when we consider the overwhelming task of improving science education, but the impact is not just in mere numbers. As of this year over 100 teachers will have gone through the workshops. They are steadily building a national network to help improve their professionalism and share their experiences. Many have gone on to win recognition for their teaching or to become influential in administration of science programs in their districts. Every year each one of them comes in contact with many students whom they can influence strongly.

Finally, we are confident we can multiply what is done in this program with our other efforts aimed at improving elementary and secondary science education through Science Alliances and at our manufacturing sites.

Literature Cited

1. *High Schools and the Changing Work Place..... The Employer's View*, National Academy of Sciences (1984).
2. *A Nation at Risk,* U.S. Excellence in Education Commission, Washington D.C.: U.S. Department of Education (1983).
3. *The Present Opportunity in Education*, Triangle Coalition for Science and Technology Education (Sept. 1988).
4. *Opportunity in Science Education*, John M. Fowler, ChemTech *19*, 79 (1989).
5. *Polymer Chemistry for High School*, National Science Teachers Association, #PB-56/1 (1989). NSTA, 1742 Connecticut Ave NW, Washington, D.C. 20009.

APPENDIX

Here is a condensed summary of the curriculum for the 1990 Workshop.

Welcome to Du Pont--Workshop orientation
A Brief History of Polymers
The Ultimate Consumer, A Slow Learner
Introduction to Polymer Chemistry and Physics
Polymer Laboratory
Science Should Be Fun

Polymer Solutions and Gels
Elastomers Laboratory
Biomedical Polymers
Rubber/Plastics Theory
Fibers Theory
Fibers Laboratory

Fibers Tour
Polymer Composites
Polymer Processing Tour
Polymer Characterization
Engineering Plastics
Physical Testing Laboratory Tour
Polymeric Waste and Recycling

Chemistry and Biotechnology
Introduction to Agribiochem
Introduction to Cloning Laboratory
Cloning Laboratory

Cloning Laboratory (II)
Tour of plant tissue culture and transformation facility
Applied Biotechnology, Analytical Systems
Tour of Stine Farm. Demonstration of Transgenic Plants
Welcome to Engineering R & D
Introduction to Bioengineering
Introduction to Laboratory Experiments and Procedures
Quantitation of Bacteria--Laboratory
Qualitative Analysis of Bacteria--Laboratory
Microbial Kinetics--Laboratory
Laboratory Review and Discussion

Introduction to Laboratory Exercises
Counting Plates--Laboratory
Examination of Selective Plates--Laboratory

Introduction to Biomaterials
Biomaterials--Laboratory Demonstration
Microscopy--Laboratory Demonstration
Automated Growth Monitoring--Laboratory Demonstration
Discussion of Laboratory Results
Review of Group Results and Wrap-up
Tour of Franklin Institute, Philadelphia

Utilization and Curriculum Planning Session
Tour of Chestnut Run Technical Service Facility
Farewell Banquet and Awarding of Certificates

RECEIVED April 11, 1991

Chapter 7

Science Education Initiatives in the University

G. A. Crosby and J. L. Crosby

Science, Mathematics, and Engineering Education Center, Department of Chemistry, Washington State University, Pullman, WA 99164–4630

A series of programs developed at Washington State University to attack some of the manifest problems in science education is described. Brief accounts of the operation of a summer science camp for eighth through tenth grade students, a summer in-service program for high school teachers of chemistry, an electronic bulletin board to network the chemistry teachers, and a new Master of Arts in Chemistry degree are presented. The latter program, designed expressly for those who are currently teaching chemistry but who do not possess degrees in the discipline, is supported by State, Federal, and private funding. Descriptions are also included of a summer program for middle school teachers that involved both the chemistry and physics departments and of a pilot tutoring program in chemistry for underprepared college freshmen. Current developments and plans for a new required science sequence for elementary education majors are outlined.

Science education and science literacy are national concerns. Insufficient numbers of talented students are choosing careers in the quantitative sciences, and the comprehension of science principles and issues by the public is far below that deemed adequate in a democracy. Although these trends become manifest at the university, the genesis of the situation lies in the precollege years, perhaps even in the elementary grades. At Washington State University a series of programs has been initiated to address these issues. Efforts began as long ago as 1983, but activity has been increasing in recent years and new projects are slated for execution in the near future. In this chapter we describe several of the programs in enough detail to give the reader some guidance should he or she be inclined to initiate similar activities on his or her home campus.

Programs for Precollege Students

Cougar Summer Science Camp. The Cougar Summer Science Camp (CS_2C) was launched in 1983 with a starter grant of $10,000 from the Washington State University Foundation. The purpose was to introduce precollege students to science and university life. Initially designed to attract those who had completed the ninth and

0097–6156/92/0478–0066$06.00/0

tenth grades, the program was opened in 1985 to students who had just completed the eighth grade. The Camp started with a contingent of 37 students and has grown to a steady-state number of ca. 110-125 per year. During the 1990 session the group consisted of 66 eighth graders, 25 ninth graders, and 21 tenth graders. Older students are not admitted, since there is insufficient time remaining for them to modify their high school programs.

CS_2C is a week-long event. Students arrive on Sunday afternoon and are discharged on the following Saturday at noon. The intervening time is packed with structured events. Little spare time remains for shopping, swimming, and other unstructured activities, but the tight schedule prevents the incidence of discipline problems. All Campers are *required* to attend *all* events. No absenteeism is tolerated. Aside from the grade restrictions and mandatory attendance, the other rigid rules are that no participant is permitted to bring a car to campus and a Camper must participate in the graduation exercises to receive a certificate of completion.

The avowed purpose of CS_2C is to demonstrate to precollege level students that science is not boring, abstruse, nor impossible for the average student to understand and appreciate. The mechanism is to immerse the students in science-oriented activities throughout the week.

A quick perusal of a typical daily schedule (Figure 1) shows that the students are kept busy from early morning until sometimes quite late at night. Also the schedule reveals the variety in the program. Lectures with demonstrations, extensive hands-on laboratory activities, computer exercises and science films are interspersed with talks by guest lecturers from other fields, tours of university facilities, visits to points of interest, and social events.

6:30 AM	Wake-up Call
7:00 AM	Breakfast
7:40 AM	Lecture/Demonstrations
9:00 AM	Computer Class/Lab
10:00 AM	Tour of WSU Radio/TV Studios*
11:00 AM	Tour of Hydraulic Labs*
12:00 N	Lunch
1:00 PM	Lab Tutorial
1:30 PM	Hands-on Laboratory (Chemical & Physical Principles)
4:00 PM	Free Time**
5:00 PM	Dinner
	Evening Programs (a field trip*, a guest lecturer, a night at the theater, a dance, etc.)
10:30 PM	Floor Check
11:00 PM	Room Check/Lights OUT

* During the week the Campers visit 17 units on campus. Other field trips include the College of Veterinary Medicine, greenhouses, physics and geology displays, dairy barns and beef cattle center, anthropology and entomology museums, anatomy labs, planetarium, observatory, nuclear reactor center, backstage of the theater. . .

** During Free Time swimming, tennis, basketball, volleyball, etc. are available.

Figure 1. Typical daily schedule for participants in Cougar Summer Science Camp.

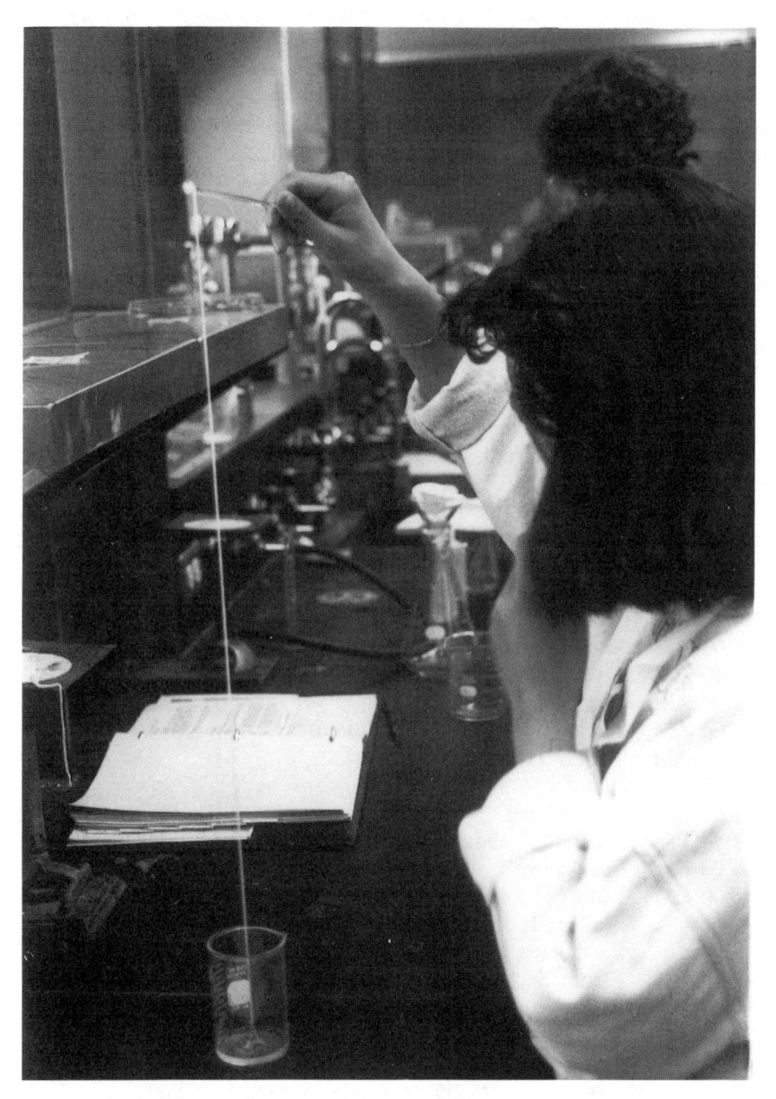

A Science Camp participant pulling a 'nylon' thread.

The week closes with the main social events consisting of a picnic followed by a dance, and then the entire program is capped with graduation exercises and a chemical spectacular---an hour of demonstrations in which colors are produced by as many reactions as possible. This final event is attended by the Campers, their siblings and parents, or friends who have come to take the participants home. The week literally ends with a ($H_2 + O_2$) bang!

Although an all-inclusive fee is charged ($300 for 1991), the Camp is not self-supporting (some subsidy from the University is necessary). More than half of the fee is required for room and board, liability insurance, medical insurance, Camp T-shirts, and Camp photographs. The remainder of the fee helps to support the cost of chemicals and supplies, incidental charges such as film rentals and extra lifeguards, and the staff. Chemistry graduate students assist in the lab, stay with the students in the dorms, accompany their assigned groups on the tours to maintain order, and serve as counselors and confidants on a 24-hour basis. There is one graduate assistant for each 25 students. In addition the director who is also the major lecturer, the program manager, and the two computer lab instructors also live in the dormitory. Each summer four to six Camp Alumni volunteers are selected to serve as Camp Aides (after their junior or senior high school year). The pace is grueling; the staff averages only three or four hours of sleep per night.

Who attends CS_2C? Posters, brochures, and application forms for the Camp are sent to all schools in the State where potential applicants are enrolled. In addition, the Camp is advertised in the alumni newsletter, on the local TV-cable channel, and on the University radio station. [The latter two broadcast through translator stations to many parts of the region.] The intent is to serve the citizens of Washington, but students do apply from other states, and a few have attended from as far away as Massachusetts, Hawaii, and even Hong Kong!

Since no fellowships are available through the Camp, there is an enrollment limitation imposed by the fee. Nonetheless, students from indigent families are supported in numerous ways---by local service clubs, by their school districts, and by corporations and private donors. Over the years the percentage of minority students and students from economically depressed areas has risen and now amounts to 30-40% of the total enrollment. Currently there are no academic criteria invoked to select the participants. Considerable self-selection occurs, however, since many of the students are oriented toward science, and the roster of fellowship students is skewed toward those with interest in science because they are recommended by their teachers.

What is the principal message conveyed by CS_2C to the students and their parents? Throughout the week the students are shown in many subtle ways what kinds of opportunities and careers there are for those who achieve higher levels of education. The CS_2C message is: Do not waste those precious high school years---a good high school background in English, foreign languages, science, and mathematics is a prerequisite for success in later years. The students receive this message continually throughout the week, sometimes explicitly, but always subliminally.

Is CS_2C successful? This is a difficult question to answer. The students start the week a little frightened. They are hesitant, diffident, and reserved. They end it with irrepressible enthusiasm. Many do take the message of the Camp to heart and their letters to the staff confirm that their eyes were opened and that CS_2C has made a difference in the way they approach their remaining high school years. Quite a few of them show up later on the campus as freshmen, and some attribute their selection of school and career choice to the experience of the Cougar Summer Science Camp.

Advanced Cougar Summer Science Camp. Numerous requests from both Camp alumni and their parents have prompted us to consider offering an Advanced Camp commencing in 1992. The philosophy of operation of this camp will be, however, different from that of CS_2C. Whereas no academic selection criteria are imposed for admission to CS_2C, the opportunity for attending the advanced program will be determined by the progress of the student in school, particularly performance in math and science courses. Moreover, funds are being sought to defray the cost of attendance for those from indigent families. The authors believe that the participants in a second camp should be exposed to more science for a longer period of time, and that they should be instructed in a more formal and demanding manner than they experienced during their first exposure to the University. Graduates of this second camp would be prime candidates for programs where the emphasis is on research and the student is employed for most of the summer. Of course, participants in such programs would be more mature.

Programs for High School Teachers of Chemistry

Northwest Regional Leadership Workshop for High School Teachers of Chemistry. During the month of July 1987 the authors held a Teacher Enhancement Workshop for High School Teachers of Chemistry in the Northwest (NWW). Supported by the National Science Foundation 27 teachers from Washington, Oregon, Idaho, and Montana were brought to the campus for intensive instruction in chemistry. The focus was on leadership and the candidates were chosen for leadership potential. The intent was to assemble a cadre of professionals who would catalyze further improvements in the teaching of chemistry throughout the region. The subsequent record of the teachers shows conclusively that the goal of NWW was achieved.

The NWW schedule was a busy one for the staff and for the participants. Not only were the days filled with lectures, labs, and computer instruction, but the evenings were often filled as well. To assure that the material presented was available for future reference, an experienced notetaker took notes in all the lectures, and edited, professional copies were distributed to all participants. Moreover, the teachers had photocopying privileges and were encouraged to use this facility to increase their supply of demonstrations, lab procedures, and public domain computer programs. All made good use of the opportunities.

Selection of the teachers was not only dependent upon the qualifications of the individuals, but also on the commitment of each local district to support its teacher in the endeavor. Acceptance into the program was contingent upon the district's supporting the teacher to return to the WSU campus for two weekends during the following school year (November and March) when the teacher was asked to report how the knowledge and experience gained during the first summer's program was being put to use in his or her chemistry program. The rationale was a simple one: If the district would not support the teacher, then the expenditure of funds by the NSF to train the teacher would not be appropriately leveraged. Returning twice during the year was exceedingly beneficial. Not only did the teachers gain from hearing of the experiences of their peers, but they also gained from the very act of reporting, both orally and in writing, to the group. For some of them this was a first experience of that kind.

Funds were obtained to bring the teachers back to the campus for a 10-day period during the following summer. Fifteen of them were able to accept the invitation. The purpose of this second phase of the program was to develop outreach activities that could be exported to other districts and other regions. The group spent two days defining the kinds of workshops that would be most valuable to other teachers, and then small groups began to formulate workshops. "Reactions for All Reasons" and "The Use of Computers in the Teaching of Chemistry" were chosen for refinement and eventually put into a form sufficiently detailed to allow a group to stage workshops with these themes.

Reaching the goal of constructing a quality workshop to be offered to one's peers turned out to be a strenuous exercise. By the end of the 10-day period the frameworks of both workshops were completed---proposed hourly schedules and lists of workshop requirements were assembled---but the written materials to be distributed to workshop attendees were still in unacceptable condition. These were available only after several months of rewriting and editing by the director and staff. The materials still are continually being updated and revised by both the teachers and the former NWW staff for use by the teams in their outreach workshops and demonstrations.

The outreach efforts of the original group of NWW trained teachers have been many, intensive, and effective. Members of the group have given numerous presentations, usually as small teams, at local school districts, at statewide meetings of science teachers, and at national meetings (ACS, NSTA). Moreover, four of the group comprised the principal members of the team that ran OPERATION PROGRESS I, an NSF-sponsored teacher-enhancement program at the 11th Biennial Conference on Chemical Education held at Atlanta, Georgia in the summer of 1990.

The evident success of NWW in developing the leadership potential of high school teachers of chemistry rests firmly on four elements of the program: (a) Interactions among the teachers themselves and with the NWW staff have been maintained over a long period of time (years). (b) Preparation and quality of performance were stressed throughout. (c) Good, well-illustrated written materials were assembled and edited before distribution. (d) A professional attitude and a high level of performance were demanded by the Director at all times. If any one of these features had been absent, the program would have been seriously impaired and the objectives not realized.

Electronic Bulletin Board: Networking Teachers. Teachers are often isolated, both geographically and intellectually. The chemistry teacher is frequently the sole exponent of that discipline in the school, or possibly the district, and professional stimulation is obtained only at an occasional meeting. Teachers need a network, not only to reduce their sense of isolation, but to allow them to access information, trade ideas with their peers and, in general, to participate in the fraternity of professionals in their discipline.

The construction of such a network is underway with the support of a grant from the NSF. An Electronic Bulletin Board (EBB), housed in the Chemistry Department at Washington State University, has been designed to serve the teachers of chemistry within the State with options to extend the service to other regions, and to other science disciplines, in the future.

The project rests on the assumptions that (a) teachers are busy and will not use a board unless there is exciting, relevant, and useful material on it, (b) a board must be free of glitches, kept up-to-date, and be easy to use, (c) the board must be accessible with no charge to the teacher or to the district, and (d) teachers will be more willing to try something new when personal instruction on the use of the board at his or her own site is offered. All of these features are designed into the EBB to serve chemistry teachers in the State of Washington and in the Northwest.

It is the goal of the EBB to keep the busy secondary teacher abreast of the happenings in the world of chemistry and chemical education. The Board provides the teachers with an easy way to ask technical questions of experts in that field, to communicate with each other, to share their classroom successes and problems, to exchange ideas, to list for barter, purchase, or sale, such items as excess chemicals, equipment, and books. The potential uses for the EBB are endless. The Main Menu of the current board is reproduced in Figure 2.

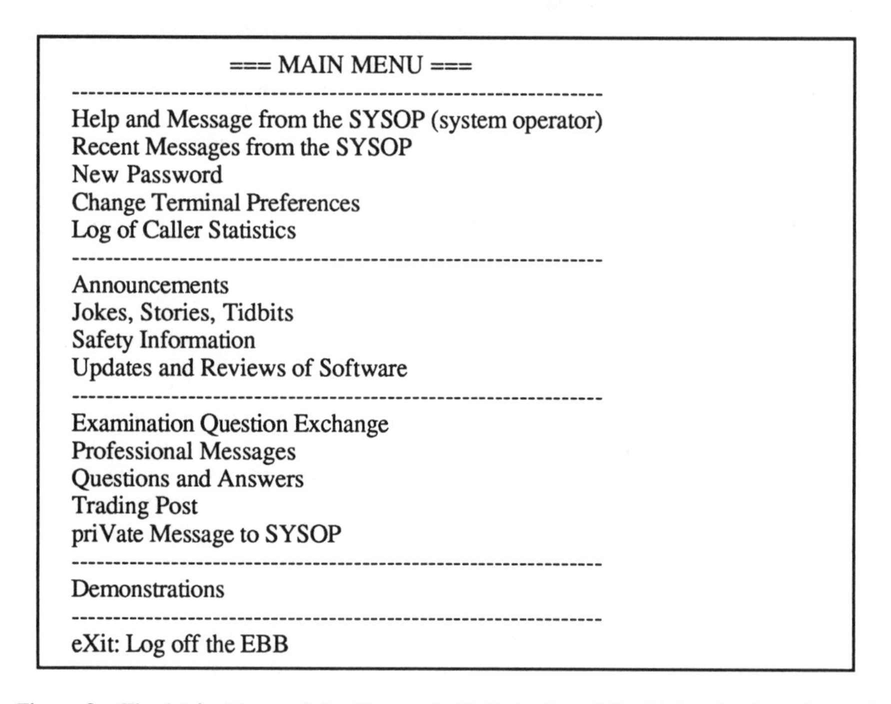

=== MAIN MENU ===

Help and Message from the SYSOP (system operator)
Recent Messages from the SYSOP
New Password
Change Terminal Preferences
Log of Caller Statistics

Announcements
Jokes, Stories, Tidbits
Safety Information
Updates and Reviews of Software

Examination Question Exchange
Professional Messages
Questions and Answers
Trading Post
priVate Message to SYSOP

Demonstrations

----

eXit: Log off the EBB

Figure 2. The Main Menu of the Electronic Bulletin Board for high school teachers of chemistry.

Each menu selection has a submenu, and some have sub-submenus. By pressing the first capital letter of any selection, the caller enters a submenu. The submenu for Announcements is found by pressing "A" (Figure 3).

A system operator must monitor the EBB daily to maintain its integrity. Moreover, the operator must also be technically competent in chemistry and have access to consultants to answer queries from the teachers. Finally, a significant continual effort must be made to add new information to the EBB regularly and to delete outdated material. Our estimate is that the equivalent of one day per week will be necessary to maintain the EBB once it is well-established.

The hardware needed to communicate with the EBB is unsophisticated. Most schools already have a computer that can communicate with a host through a modem. The EBB server is a Macintosh, and thus a Macintosh at the school is ideal since detailed graphics can be downloaded and printed. The EBB can communicate with IBM computers or clones, Apple II machines, and others, since most of these computers have inexpensive software packages that allow them to talk to host machines. On the EBB all files that are designed for downloading are supplied in several versions in order to accommodate most types of computers. In some instances graphics information may be lost, but text always comes through faithfully.

==== Announcements ====

Awards, Grants, and Fellowships
ChemCom (Chemistry in the Community)
Degree Programs
Events in Science: Mark Your Calendar
Free Literature and Publications
Meetings and Conferences
Opportunities and Programs for High School Students
Professional Memberships and Journals
Recent Publications
Workshops and Institutes
eXit to Main Menu

Figure 3. A copy of the Announcements submenu from the Electronic Bulletin Board.

To assure that a critical mass of teachers would be reached and thus make the EBB cost effective, funds were obtained to support an experienced teacher/user to tour the entire state and other parts of the Northwest to show his peers how to access the EBB. This operation is currently underway. Moreover, this EBB expert also carries a supply of modems and cables to fit various types of computers. These items, obtained through a discount supplier that does not accept purchase orders, are sold to the school districts at cost. The service has turned out to be quite valuable, particularly to rural schools that do not have local electronics or computer stores.

Currently, over 160 teachers mainly in the State of Washington have been validated to use the EBB. After the one-semester visitation program is completed (June 1991), the number of validated users is projected to exceed 200. This number should increase with time as knowledge of the EBB and its usefulness spreads among teachers. Invitations to become a participant in the project will also be mailed periodically to nonparticipating districts over the next year.

The success of any electronic bulletin board can only be measured by the number of individuals who use it and the amount of time they spend accessing it. Success will also be evident if the number of users increases with time. Fortunately, the groups of teachers who have attended workshops at WSU already provide a basis for building an infrastructure.

Master of Arts in Chemistry for Practicing Teachers. A pandemic problem in the public schools is the incidence of teachers teaching out-of-field. Particularly in rural districts, an individual with a degree in one science, such as biology, may be forced to teach chemistry, physics, or even math and computer science. Since the teacher usually has minimal preparation in these ancillary areas, classroom performance is often below acceptable standards. In particular we address the teaching of high school chemistry that is frequently assigned to teachers with minimal training in the field, particularly in quantitative chemistry. Most of the teachers possess insufficient education to be able to run a laboratory-centered high school chemistry course competently and safely.

Often the State demands that a teacher attain a Master's Degree within a fixed number of years of service and usually the pay scale is tied to that attainment. Summers are available for teachers to acquire additional education, but summer programs in the sciences that meet the needs of the practicing teacher are not usually offered. For instance, a Master of Science degree in chemistry is generally designed for students preparing for industrial employment or further graduate study. Furthermore, these programs presuppose an undergraduate major in chemistry---something the practicing high school teacher does not normally possess. The question is clear. What kind of program should be designed for the out-of-field practicing chemistry teacher that will best serve the needs of the individual, the school, and the public? It behooves the university chemistry departments to construct a graduate course program that possesses integrity, leads to a higher degree, and improves the knowledge and competence level of the teacher for managing high school programs in chemistry.

At Washington State University a new degree, the Master of Arts in Chemistry, has been designed and is being implemented. Crafted explicitly for out-of-field teachers, the proposed program is scheduled to be completed in three years. The program is set; there are no options. Two special features are: (a) Instruction during the academic year is carried out via the use of VCRs (videotape by mail). (b) The program includes a two-month summer assignment at a Federal laboratory (or nearby site) with concomitant instruction via two-way interactive television from the home campus. Descriptions of all the newly-designed courses can be found in Appendix A.

The course outline for this degree reveals why it is neither a Master of Science in Chemistry nor a Master of Arts in the Teaching of Chemistry. It clearly is not focussed on the frontiers of chemical knowledge as required of the former and it does not emphasize pedagogy to the extent that the latter degree usually does. Hence, the degree is classified as a Master of Arts in Chemistry.

Two other features of the course structure deserve comment. First, there is a heavy emphasis on laboratory work, since high school teachers of chemistry are often insufficiently prepared to run labs safely, economically, and efficiently. Second, there is a decided thrust of all the courses in the direction of practical chemistry (demonstrations, molecular basis of devices, consumer chemicals) to enable the teacher to teach relevant chemistry in a less-than-optimal environment. The intent is to equip the teacher to solve the real problems of teaching in a setting where equipment is primitive, support is minimal, facilities are inadequate, and professional isolation prevails.

The capstone assignment in a non-academic laboratory is a unique feature of this degree program. Not only will the teacher earn a reasonable salary during the two-month appointment at the laboratory, but he or she will be a member of a scientific team that is engaged daily in the solving of chemical problems. Since the emphasis of these particular collaborating laboratories is on environmental problems (and hence chemical analysis plays an important role), the degree course program also features analytical methods and methodology to enable the teacher to be a valued contributor during the summer program at the Federal or corporate lab.

The funding for this degree program is shared by the NSF, the private sector, the State of Washington, and the Department of Energy. The program uses Federal and private facilities in a productive way to educate practicing teachers in conjunction with a university degree program.

Programs for Elementary and Middle School Teachers

Middle School Physical Science Workshop. There is a growing national perception that the roots of the problems in science education trace back to the middle school or even to the elementary years of schooling. Statistics reveal that the amount of time spent on science in the elementary classroom is far less than that spent on other subjects, such as reading. Part of the problem stems from the education and training of the teachers, since many of them have poor science backgrounds, particularly in the areas of elementary principles of physics and chemistry. Thus, they are uncomfortable teaching science concepts and tend to slight these subjects in their daily activities. In fact, the insecurity of the teachers in the areas of physical science may be partially responsible for the aversion to science that begins to manifest itself in children before middle school.

To improve the science backgrounds of practicing middle school teachers a program was run jointly by the Departments of Chemistry and Physics at WSU in the summer of 1989. Financed through the Office of the Superintendent of Public Instruction the program, Explorations in Middle School Science, was attended by 27 fifth- through ninth-grade teachers. The first two weeks were devoted to physics concepts, both lecture and laboratory, and the latter two weeks were spent on elementary chemistry. Wherever possible simple equipment and materials that could be found in local hardware stores or supermarkets were used. Detailed notes of all lectures and illustrated writeups of the experiments were provided to the participants in the hope that the teachers would use some of the activities as demonstrations, or possibly as student activities, in their classrooms.

The results of the program were mixed. Since some of the ninth grade teachers had reasonable science backgrounds, but most of the fifth and sixth grade teachers did not, some tension developed, particularly in the labs. All learned a great amount, as revealed by exams and questionnaires, and a few of the lower-level teachers profited immensely, at least according to their submitted evaluations. Most of the participants thought that the program should have focussed on physics for one three-week term and then on chemistry for a second three-week term during the following summer. Most believed that the program was too compressed and overly challenging for the short time allotted.

From this first attempt to improve the teaching of prehigh school science through an inservice program, several conclusions were drawn. First, the poor, sometimes nonexistent, science backgrounds of elementary teachers must be recognized. Although they are eager to learn, many of the teachers harbor misconceptions about basic physical phenomena such as density, heat and temperature, and vapor pressure. They must experience these phenomena in the laboratory, first-hand, or verbal explanations are meaningless. Second, quantitative reasoning, even simple ideas of ratio and proportion, must be reinforced daily. Finally, the realities of the middle school classroom must be factored into the inservice program. Chemistry without sinks, without Bunsen burners, without balances, and without storage facilities is the norm for the elementary/middle school classroom. To run an effective inservice activity the university instructor must come to terms with this situation or the instructional program will never be translated into practice in the middle school classroom.

Much must and can be done in this area of education. If the stated goals of making the U.S. first in science and mathematics education by the end of this century are to be met, a concerted effort by the university science departments to become involved in teacher inservice programs will be necessary. Joint efforts that establish links among science departments, the education schools, and the school districts are especially needed.

Preservice Science Programs for Prospective Elementary Teachers. To prepare for the 21st century a new paradigm must emerge for the preservice programs for elementary teachers, a paradigm that recognizes the centrality of science and mathematics in the modern world and establishes these subjects at the core of the educational programs for prospective teachers. To achieve the goal of good science education for all students, the current curricula for elementary teachers must be drastically restructured. Without backgrounds in *relevant* science the elementary teacher will continue to avoid teaching science or, when forced to do so, communicate a distaste and possibly a fear of the subject to the students. Reform efforts must involve both the Colleges of Education and the Colleges of Sciences and Arts since the former have the responsibility for teaching pedagogical skills and the methods of handling children, whereas the latter are the repositories of science knowledge. At WSU efforts to increase the science components of the elementary education program have been underway since 1987. Supported by a grant from NSF, two new courses were added to the elementary education curriculum: a course in Astronomy/Physics and a course in Chemistry/Earth Sciences. The former is taught in the Physics Department and the latter in the Chemistry and Geology Departments. Both are semester (4-hr) courses and each includes a lab. The courses are still evolving, but the general outlines are well defined.

Although these new courses are science courses, they are taught in a manner different from the usual first courses taken by undergraduates. In the chemistry segment (1/2 semester) the lectures are augmented by copious demonstrations, and inquiry methods are used to elicit student response and involvement. The labs are also designed around consumer chemicals, commonly available supplies (supermarket items), and easily obtainable equipment. The intent is to engage the prospective teacher intellectually during the lecture using pedagogical methods that are particularly appropriate to an elementary classroom. The hope is that these students will also teach science in this manner when they are in control of their own classrooms.

Experience with students enrolled in the chemistry course who had not yet taken any mathematics at the university revealed that a mathematics prerequisite (algebra) was necessary. This requirement is currently being invoked. Moreover, although the students enjoy the labs, they retain little of value *unless* they are required at the end of the lab sessions to answer questions, in written form, about the experiments. Finally, the students possess utilitarian attitudes. They want to learn only that which will be of immediate use in the elementary classroom. This translates into doing lab experiments that can be used later as classroom demonstrations, relating principles immediately to everyday phenomena, and focussing on the very basics of chemistry. A partial list of the laboratory activities is reproduced in Figure 4 to show the tenor of the chemistry section.

Although these science courses designed explicitly for prospective elementary teachers are deemed to be successful, the proposal is to increase the amount of science in the elementary education program by expanding the science component still further. Under the plan the half-semester of chemistry will increase to a full semester, and a new course in Geology/Ecology will be introduced. Eventually, a biology sequence designed explicitly for the elementary education curriculum will be added. In cooperation with the College of Education plans are underway to make science the focus of the elementary education program. If this program is fully implemented new teachers should emerge from the university both ready and anxious to teach science in a manner that will evoke the natural curiosity and wonder of children.

Preparation of Cabbage Juice Indicator
Preparation of 2 M Sodium Hydroxide (from Red Devil™ Lye or Drano™)
Preparation of Phenolphthalein Indicator (using Ex-Lax™)
pH Calibration of Cabbage Juice Indicator
pH Determination of Some Household Products
Acid/Base Chemistry (with consumer products)
Vapor Pressure vs Temperature: "Pop-Can Crush"
Boyle's Law: "Cartesian Diver"
Endothermic/Exothermic Reactions: "Cold Pack/Hot Pack"
Thermal Conductivity: "Boiling Water in a Paper Cup"
Generation and Properties of Carbon Dioxide (from washing soda and vinegar)
Generation and Properties of Oxygen (from yeast and peroxide)
Generation and Properties of Hydrogen (from steel wool and muriatic acid)

Figure 4. A listing of typical experiments performed by elementary education majors in the special half-semester chemistry course.

Program for College Students

Tutoring for Retention. Tutoring centers exist on most campuses. These are often housed in minority centers and are not necessarily staffed by tutors competent in the sciences. In many cases these tutors are not selected by the faculty from the specific disciplines. At WSU we have installed a tutoring program for general chemistry incorporating several unique features designed to correct some of these perceived deficiencies. The program is experimental and is based on the following premises: (a) Many of the difficulties experienced by students in math and science courses stem from their basic lack of appropriate attitudes and habits for mastering quantitative subjects. (b) Often students focus on the inessentials of a course and rely on memory rather than reasoning to conquer a topic. (c) Many students need a peer-group support structure to help carry them through a difficult course.

Although this tutoring program at WSU is voluntary, *attendance is mandatory*, i.e., once enrolled, a student must continue to attend or be disenrolled from the program and barred from returning. Specifically, an earnest fee of $25 is charged that is returned to the student *only* if the student faithfully attends the tutoring sessions for the entire semester. (If the student drops the regular lecture course, however, the fee is returned.) The program is time-consuming for the student since a tutoring session is held after each class lecture, usually in the late afternoon of the same day. In a real sense the program forces the student to spend time on the subject before the echoes of the lecture have receded too far from memory.

The tutor is carefully chosen not only for knowledge of the subject but also for skill in exposition. Moreover, the tutor attends the class lectures, takes notes, and distributes edited copies of these notes to the students enrolled in the tutoring sessions. This helps to validate the notes that the students have themselves taken in class. The tutor also issues additional problem sets, designs practice tests to prepare the students for forthcoming exams, and often fills in background material that the students need to comprehend the regular lectures. (Converting to exponential notation, solving simple equations, and manipulating logarithms are examples of such exercises.)

This pilot tutoring program carries with it no grade. The students attend the regular class sessions, take the class exams with the other students, and are graded by the instructor along with the rest of the students. In fact, except for the presence of the tutor taking notes, the class instructor could be entirely unaware of the program.

Is the tutoring program successful? Average grades earned by these students on the class exams are consistently higher than the class averages, thus signaling success. Since most of the students in the tutoring program entered the University with deficient high school backgrounds, it is probable that they would have fared much worse had they not been part of the program. The students who stay with the program serve as enthusiastic recruiters for it, which is a sign of their confidence in the worth of the sessions. At least one can say that the program is fiscally responsible, since the students attend the sessions and keep up with the work. The earnest fee appears to be a sufficient incentive. Thus, the usual practice of students ignoring tutors until just before exams is eliminated, and the salary paid to the tutor is indeed earned throughout the semester. The program is presently under evaluation and consideration for expansion to other science classes.

Summary

Many initiatives are being taken by universities across the Nation to improve educational opportunities for precollege students, for practicing and prospective teachers, and for undergraduates. In this chapter we have described a few programs undertaken at Washington State University over the last decade. The programs have been designed to solve particular problems within the constraints of the institution and the region. Some of them have potential for export to other institutions and other regions. Nevertheless, they can only be considered as guides for thought and program development, since education in the U.S. is a local and state issue and a program that works well in one environment may not be suitable for another. Nonetheless, the authors are convinced that, in spite of the idiosyncratic nature of many educational practices, the programs described here do address some of the major problems plaguing the Nation in the area of science education.

Appendix A: Description of Courses for Master of Arts in Chemistry

Courses marked with an asterisk (*) will be offered via videotape during the academic years.

CHEM 411: General Chemistry from an Advanced Point of View

Quantitative aspects of chemistry; first law of thermodynamics, solution theory, equilibrium, kinetics; electrochemistry and redox reactions; inquiry and problem solving.

CHEM 413: Lab Preparations, Methods, and Management (Lab course)

Synthesis, analysis and reactivity; reactions and methods appropriate for high school; microscale chemistry; time-saving techniques, inventory control, safety and disposal.

CHEM 416: Lecture Demonstrations and Their Uses (Lab course)

Purposes of lecture demonstrations; multiple uses of demonstrations as a tool for motivating students and using inquiry techniques; safety and disposal.

CHEM 419: Physical Foundations of General Chemistry

Basic principles of physics underlying general and biophysical chemistry.

CHEM 505: Molecular Basis of Modern Materials and Devices*

Atomic and molecular structure; the solid state; materials science; transition metals and coordination complexes.

CHEM 571: Organic and Biochemistry I*

Organic structures and functional groups; reactions; consumer chemicals; polymers, biopolymers and macromolecules.

CHEM 572: Organic and Biochemistry II*

Continuation of Organic and Biochemistry I with emphasis on biochemical structure and function.

ED ADM 520: Seminar in Curriculum and Instruction*

Analysis of high school science and math curricula; teaching and inquiry techniques; implications of modern research results for science teaching.

CHEM 519: Analytical Methods and Instrumentation

Principles of modern analytical methods; separation techniques; trace analyses.

CHEM 575: Survey of Biophysical Chemistry

Survey of applications of physical chemistry to molecular biology; thermodynamics, solutions, electrochemistry, phase equilibria, kinetics, transport and spectroscopy.

CHEM 506: Industrial Practicum

Assignment in an industrial laboratory.

CHEM 702: Master's Special Problems

Design of a modern high school chemistry course; lecture outline; demonstrations; laboratory methods; inventory control; equipment needs; examinations; safety and disposal.

RECEIVED May 13, 1991

Chapter 8

Technology in Science and Mathematics Curriculum

An Industry–University–School Collaboration

Paul S. Markovits[1] and Carole P. Mitchener[2]

[1]Mathematics and Science Education Center, 8001 Natural Bridge Road, St. Louis, MO 63121
[2]School of Education, DePaul University, Chicago, IL 60614

The Technology in Context (TIC) Project is one of several on-going programs for teacher professional development supported by the Mathematics and Science Education Center of St. Louis. The project involves teachers in the development of a Science/Technology/Society (STS) curriculum based on the knowledge they gain from a summer internship experience at the McDonnell Douglas Corp., St. Louis. The purpose of a three year study of the program is to develop a teacher enhancement model which does five things: (1) encourages teachers to intentionally reflect on how they teach; (2) broadens teacher's societal perspectives of technology and how technology interrelates with mathematics and the sciences; (3) develops teachers' abilities to integrate new knowledge into their curricula; (4) enhances teachers' professionalism; and (5) assesses various industry experiences for translation into classroom activities. The project model for industry, university and school collaboration involves commitment to the STS curriculum as the primary objective of the internship. Support given by industry, school districts and the university communities for implementation and dissemination of the curriculum is vital for the infusion of the curriculum into the classroom.

Scientific literacy in the United States during the decades of the 70's and 80's has declined dramatically in comparison with other industrialized nations. This decline jeopardizes our economic, social and scientific status within the world.

0097–6156/92/0478–0080$06.00/0

Concerns are being voiced from a broad spectrum of individuals and organizations. Calls for educational reform have come from the private sector, legislative bodies and national scientific and education organizations such as the AAAS (*1-2*), NCTM (*3*) and NSTA (*4*).

This project and its accompanying handbooks, "**Technology In Context: Generating Curriculum from Internships (TIC)**" (*5-6*) were developed and implemented in the St. Louis area to respond to the need for a more scientific and mathematically literate community. The project is one of several coordinated by the Mathematics and Science Education Center (MSEC) of St. Louis which addresses teachers' needs for continued professional growth in both subject area knowledge and teaching methodologies. We believe that fostering improved knowledge, methodology and dedication toward teaching will positively affect the scientific literacy of children in our schools.

Basic science, mathematics and technology are a part of scientific literacy. The TIC project focuses on technology as a ... "process by which information from science, engineering and the social fields is used to change our environment or some aspect of human existence" (*7*). The intent of the project is to help teachers incorporate "technology" as an integral facet of science education for the improvement of scientific literacy.

The Technology in Context Project

The TIC program is a collaborative effort between the private sector and education. It represents an important, growing partnership forged to improve science and math education in the schools.

The TIC model has two main components: the internship experience and the curriculum project. Those two components carry out the following objectives:

1. Provide teachers with firsthand knowledge of science, mathematics and technology in an industrial context;
2. Offer opportunities for teachers to interact with professionals from industry in the fields of science, mathematics and technology;
3. Assist teachers in translating what is learned from first-hand internship experiences into curriculum materials for use in the classroom;
4. Encourage teachers to integrate technology into the math and science curriculum using a specified orientation: a societal perspective;
5. Reinforce the advantages of viewing curriculum as an ongoing process which thrives on deliberation with colleagues;
6. Illustrate ways to integrate technology into the math and science curriculum so that others can build similar opportunities for their students, and
7. Support teachers in their professional enhancement and encourage the professional growth of other teachers.

The internship component of this project lasts for eight weeks during the summer. Teachers work on an industrial project chosen according to their area of expertise. These projects range from doing experimental laboratory work to working on computer programs. From these experiences, teachers draw out elements of technology that exemplify their work and that can be translated into student materials. Also, during the summer, teachers develop their technology curricula. During the school year, the teachers implement and revise the curriculum projects. Revision of the curriculum has culminated in curriculum handbooks from the project.

In addition to implementing and using the developed materials within a teacher's school, each teacher (intern) is required to inservice at least his/her colleagues in his/her own school and/or school district. Each teacher is also financially supported to be able to present at local, regional and/or national professional meetings.

Criteria for Selection to the Program

The TIC program is a professional development program designed to encourage teachers to learn about mathematics, science and computer science as applied to the industrial workplace. The participants are expected to act as mentors and supporters for other teachers. The uniqueness of the program is the emphases on curriculum development and presentation of materials to colleagues. The program is intended to revitalize outstanding master teachers. It also encourages younger teachers who show great potential in teaching to remain in teaching.

Selection Criteria includes:

* Demonstrated ability to write;
* Strong background in a content field;
* Involvement in curriculum development;
* Commitment to dissemination of information and techniques to others;
* Demonstrated willingness to work with others on projects, and,
* The teachers' schools willingness to support the efforts of the program.

The TIC Program (the MSEC) is not a summer employment agency. Our intent with this program at the McDonnell Douglas Corporation is to provide a valuable service to the corporation and to the educational community.

Project History

The **McDonnell Douglas Corporation (MDC)** and the **Mathematics and Science Education Center (MSEC)** formed the TIC Program as a model to support the professional development of science, mathematics and computer science teachers. The McDonnell Douglas Corporation provided summer employment for five teachers in both 1988 and 1989, and ten teachers in 1990. One individual was asked by MDC to participant for all three summers. A total of 18 teachers were placed in this project. Administrative support by Civic Progress of St. Louis (an influential local philanthropic organization) in 1988 and the McDonnell Douglas Foundation in 1989 and 1990 enabled the MSEC to plan for the development of the curriculum materials which are being used in the interns' schools and shared with other educators in the St. Louis area. The interns are also making the materials and concepts available on a regional and national level.

The support of personnel within the McDonnell Douglas Corporation was essential to the development of this project. An MDC engineer who is a Science Program Committee member for the MSEC initiated the project idea at MDC. Two corporate vice presidents added their support and a formal proposal to fund the curriculum/administration of the project was made to the McDonnell Douglas Foundation in 1987. The proposal to the Foundation was not funded due to other areas of interest at that time. However, five companies (divisions) within MDC agreed to hire interns for summer 1988 and Civic Progress of St. Louis matched a portion of the support from the Mc Donnell Douglas companies. Each company supports the interns from departmental budgets. Payment for summer employment of the interns is $500/week.

The McDonnell Douglas Foundation agreed to fund the curriculum and assessment costs for the interns during 1989 and 1990. The Foundation also announced the funding for six interns in 1991. This funding is a crucial part of the model being developed because it is used to offset costs of curriculum development and dissemination of materials associated with the internship.

The five interns in 1988 were selected from 77 applicants and the pool of candidates in 1989 was 56 for five internship positions. Nine interns who worked on curricula in 1990 were selected from 39 applicants.

Process

Following is a description of the process by which an internship is accomplished from beginning to end:

1. Job descriptions for interns are developed by McDonnell Douglas Corporation. Potential supervisors are identified. The descriptions are given to MSEC;
2. Previous summer's interns describe their programs to interested teachers, MDC personnel and others during an evening presentation;
3. An invitation to apply for the available positions is publicized to districts and schools by MSEC;
4. Teachers apply for the positions and applications are screened by a committee of the MSEC to match established criteria. Appropriate applications are forwarded to MDC companies (3-7 applications per position);
5. Applications are evaluated and applicants interviewed by McDonnell Douglas -- specifically by the potential supervisor of the teacher. The MSEC coordinates reference checks and verifies commitment of school districts of potential teachers to the project;

6. MDC makes recommendations to the MSEC. MSEC assigns interns to positions and applicants are notified. Each teacher is assigned a supervisor upon acceptance;
7. Each teacher intern attends a one day orientation meeting before beginning summer work. At this meeting, initial plans are made regarding the development of classroom materials which each teacher will take back to the classroom for use after the internship has been completed. (The 1988 interns were required to have their industry and school supervisors with them at this orientation. This has been eliminated due to the difficulty of melding schedules.);
8. The internship is completed during the summer months. During this time, weekly group meetings for each teacher and the curriculum supervisor are held to plan and discuss the STS materials to be prepared for use in the classroom. The curriculum materials are completed by the end of the summer;
9. The internship process is assessed by an outside evaluator and suggested modifications in the program are considered;
10. The completed curriculum materials are submitted to the curriculum department of the respective school systems of the teachers for discussion and approval for use in the classroom. After the materials meet approval, they are field tested and revised for inclusion in a handbook by January 1; and,
11. During the course of the following school year, teachers who have completed both the internship and the curriculum materials, along with their internship supervisors when possible, present, explain, and discuss the materials they have prepared. These presentations are at local, regional and national professional meetings. Funding for travel and materials is available depending upon the size of grant received by the MSEC.

Evaluation

Ethnographic study procedures are being used to evaluate the Technology In Context program. (*8*) Interviews with the teacher interns, their students' surveys and surveys of the McDonnell Douglas personnel directly involved with the project indicate very positive responses to the experience in 1988 and 1989.
Data for the 1990 TIC are presently being gathered.

Teachers. Teachers stated that the experience was motivating, rejuvenating, enriching, and exciting. They cited the awareness of how working in small groups was very important in industry and that they gained a great deal in knowledge base from the experience. There was a new enthusiasm to return to school and increased self-confidence. Teachers also felt that writing the curriculum made them take a closer look at their goals and objectives and analyze more carefully their lessons in all of their classes in regard to learning outcomes.

Students. Students were pleased with the "doing" science approach of the units of study. They found the lessons interesting and meaningful. The following comment from a student survey illustrates the reactions of the students:

"The best part of the exercise was the feeling of importance and accomplishment when the experience was over."

Industry Personnel. A survey of the 15 involved personnel indicated that the intern was a positive addition to their work and they felt that the teachers did a fine job and were knowledgeable in their field. Fourteen of the fifteen were interested in participating the following year and all 15 indicated that the program should be expanded.

Overall the program was perceived as being very beneficial.

Dissemination

The Technology in Context: Curriculum Handbooks, developed as a result of this project are being distributed nationally to interested educators and industry people. The handbooks uniquely tell a story about experiences teachers have had working in technological settings. The materials are based on actual work teachers did in industrial settings, and demonstrate how teachers transferred that experience to classrooms. The curriculum deals with real life technological applications, from first-hand experiences.

Over 1000 students directly used the materials developed in 1989. In 1990, over 850 teachers participated in programs using the materials. The impact from teachers who have been informed of the materials and may be using them could be greater than 15,000 students in 1990. A concerted effort is being made to accumulate and record specific use data from students in the 1990 - 1991 school year.

Programs such as this and the IISME Program from the Lawrence Hall of Science (*9*) directly impact the lives of teachers and their students. We feel that the project promotes professional development of teachers by concentrating their efforts on applications to the classroom and the self enrichment and fulfillment achieved through sharing with colleagues.

Project Variation

The extensive differences in subject matter and emphasis of each of the interns with the McDonnell Douglas program fosters individual content area applications. A variation to this MDC approach is being assessed for a project sponsored by the American Association of Immunologists at the Washington University Medical Center. The MSEC and the Biological Science Curriculum Study (BSCS) placed five interns at the Medical Center to work in individual laboratories during the summers, 1989 and 1990. A system was initiated for the 1990 interns to participate in a four week internship and to be financially supported for one week following the internship during which all five interns worked on a combined immunology curriculum package. Each intern is also supported in the field testing of the materials and in travel for presentations concerning the curriculum.

The interns viewed the experience as very beneficial and are in the process of finalizing the curriculum materials. These materials will be made available locally and nationally.

Concluding Remarks

A structured industry internship for teachers with an emphasis on curriculum is beneficial to the professional development of teachers, provides information and exciting material for children and gives industry personnel an opportunity to work on a collegial level with teachers. The formal and informal ties of industry and education which are made through these experiences provide a dimension to the classroom which impacts how students see the role of science and mathematics in their lives. The support of industry, university and professional organizations is vital to scientific and mathematical growth in our schools.

Literature Cited

1. American Association for the Advancement of Science. *Science for All Americans: A report from the Project 2061;* American Association for the Advancement of Science: Washington, D.C., 1989.
2. Champagne, A.B. and I.M. Baden. "Science Teaching: Making the System Work" *This Year in School Science 1988: Science Teaching - Making the System Work;* Champagne, A.B., Ed.; American Association for the Advancement of Science, Washington, D.C., 1988.
3. National Council of Teachers of Mathematics (NCTM). *Curriculum and Evaluation Standards for School Mathematics*; National Council of Teachers of Mathematics: Reston, VA, 1989.
4. "Position statement on science-technology-society: Science education for the 1980s; *Science Teaching: A Profession Speaks*. Brown, F.K. and Butts, D.P., Eds.; National Science Teachers Association: Washington,D.C., 1983.
5. Mitchener, C. *Technology in Context: Generating Curriculum from Internships (TIC);* Mathematics and Science Education Center: St. Louis, MO, 1990; Vol. 1.
6. Brown, C. *Technology in Context: Generating Curriculum from Internships (TIC);* Mathematics and Science Education Center: St. Louis, MO, 1991; Vol. 2
7. Hurd, P.D. *New directions in science education. Sourcebook for Science Supervisors*; National Science Teachers Association: Washington, D.C., 1987.
8. Changar, J., MDC. *Summer Teacher Internship Program Evaluation Report;* St. Louis, Missouri. 1988 and 1989.
9. Industry Initiatives for Science and Math Education (IISME). *Program Highlights*; Lawrence Hall of Science, U.C.-Berkeley, Berkeley, CA, 1988.

RECEIVED May 29, 1991

Chapter 9

University–Industry Research Partnerships

A Corporate Perspective

J. D. Burrington

BP Research, Warrensville Research Center, Cleveland, OH 44128

In the past ten years, there has been a surge in the level of industry-funded university research, driven by both the financial environment and a renewed realization of the synergistic missions of both enterprises. The nature and extent of the benefits derived will depend on the joint ability of both parties to recognize their respective expectations and to adjust key parameters which define the partnership to maximize the probability of meeting expectations. A corporate perspective on these issues and a description of the major program types funded by BP in the U.S. will be presented, including a discussion on BP's Extramural Research Award program as a means to complement and extend internal R&D programs through fundamental research.

A historical account of U.S university-industry partnerships over the past 70 years has been recently discussed (1.) and is summarized below.

Industry-funded university research of the 1920's and 30's was dominated by foundation philanthrophy, driven by the attitude that corporate funds should be used solely to benefit mankind and that the academic mission must remain uncompromised by anything but a "strict" division between its research function and the industrial laboratory. The postwar years of the 1940's and 50's saw a boom in government funding for research. These funds, however, began to encourage corporate-university alliances, while maintaining a separate academic research entity, a trend which continued into the 70's.

0097–6156/92/0478–0087$06.00/0

In the past 10 to 15 years, a significant increase, in real terms, in industrial funding for universities has occurred, nearly trebling from $330 MM in 1977 to $890 MM in 1988, and more than doubling as a percent of total university research over the period, from 3% to over 7% (Figures 1a and 1b, both figures in 1988 dollars). It is this latest surge in funding which has generated the most interest in the development of university-industry partnerships. The factors surrounding these new developments, their implications for future R&D in the U.S., and examples of how university research programs are tailored within BP for mutual benefit are presented below.

Recent Trends

The increasing level of industry-sponsored university research can be partly understood by the environment which has amplified its value to both partners. Industry has been under increasing pressure to rationalize its research and recover costs from its operating units, resulting in an intensified focus of its in-house resources on applied R&D. Universities, on the other hand, are feeling the effects of federal deficit reductions, which create the need to seek alternate sources of funding.

The result is that industry is now bearing a greater percentage of total basic research funding, increasing from 14% in 1977 to nearly 20% in 1989, an increase of $1.5 billion (in 1988 dollars, Figure 2). Industrial funding to universities as a percentage of basic industrial research has also increased over this period from 19% to 27% (Table 1). Thus, while university research is still only 4% of industry's total R&D spend, it is becoming a more significant part of its basic research portfolio.

But aside from the underlying financial environment, there is also a renewed recognition of the complementary mission of academia and industry and of the synergy that can result through collaboration. The academic mission and infrastructure, which is geared toward education and fundamental contributions to science and engineering, create an environment and a culture which is heavily leveraged by public funds for the highest quality, cost-effective basic research. Industrial participation encourages development of curricula and research which are relevant to commerical application while simultaneously providing a perspective on the critical technical challenges in industry and the need for interdisciplinary research. The increasing recognition, especially among academics, that both money and technical interaction are critical to successful collaborative programs reinforces the importance for industry to enter into a partnership with universities, and not merely function as a funding source.

Tailoring Programs

While these general benefits are recognized globally, the justification for individual programs must be based on the specific expect-

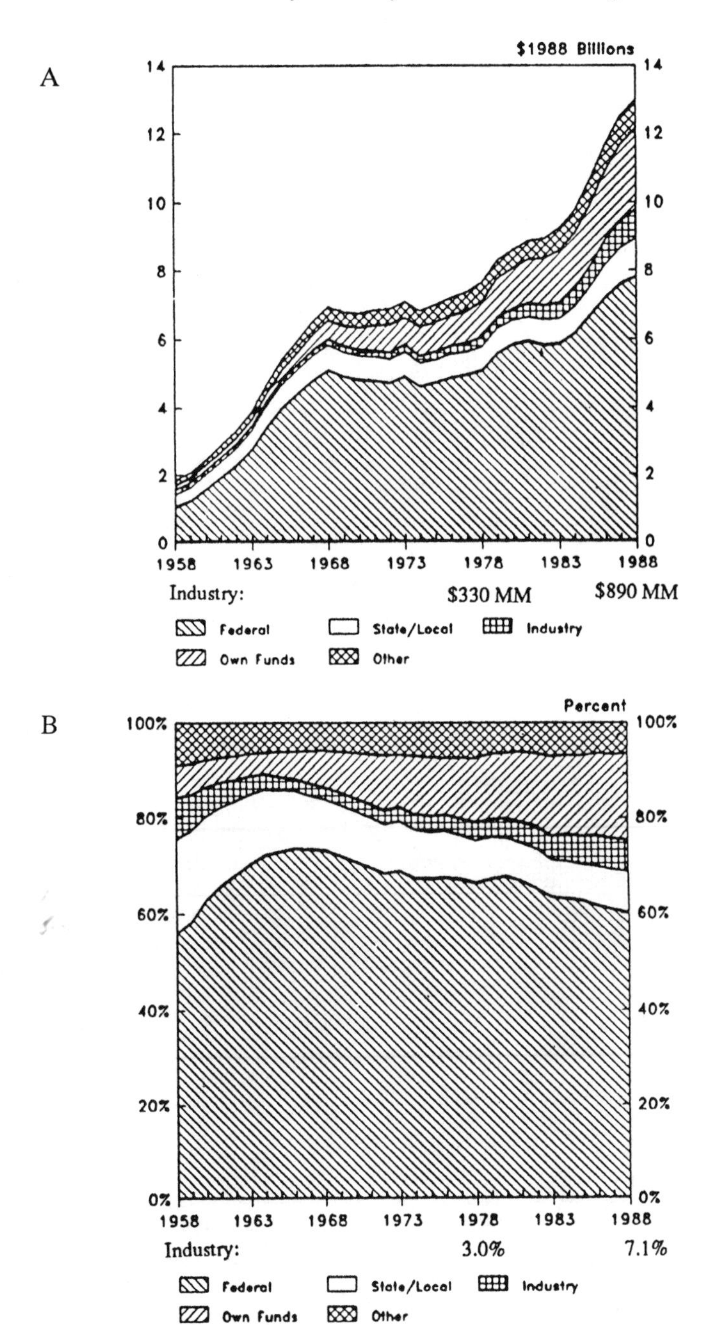

Figure 1. Academic R&D expenditures by source. *A*,Total expenditure; *B*, Distribution.

Source: "Science, Technology and the Academic Enterprise: States, Trends and Issues", The Government-University-Industry Research Roundtable, National Academy Press, Washington, DC, October, 1989.

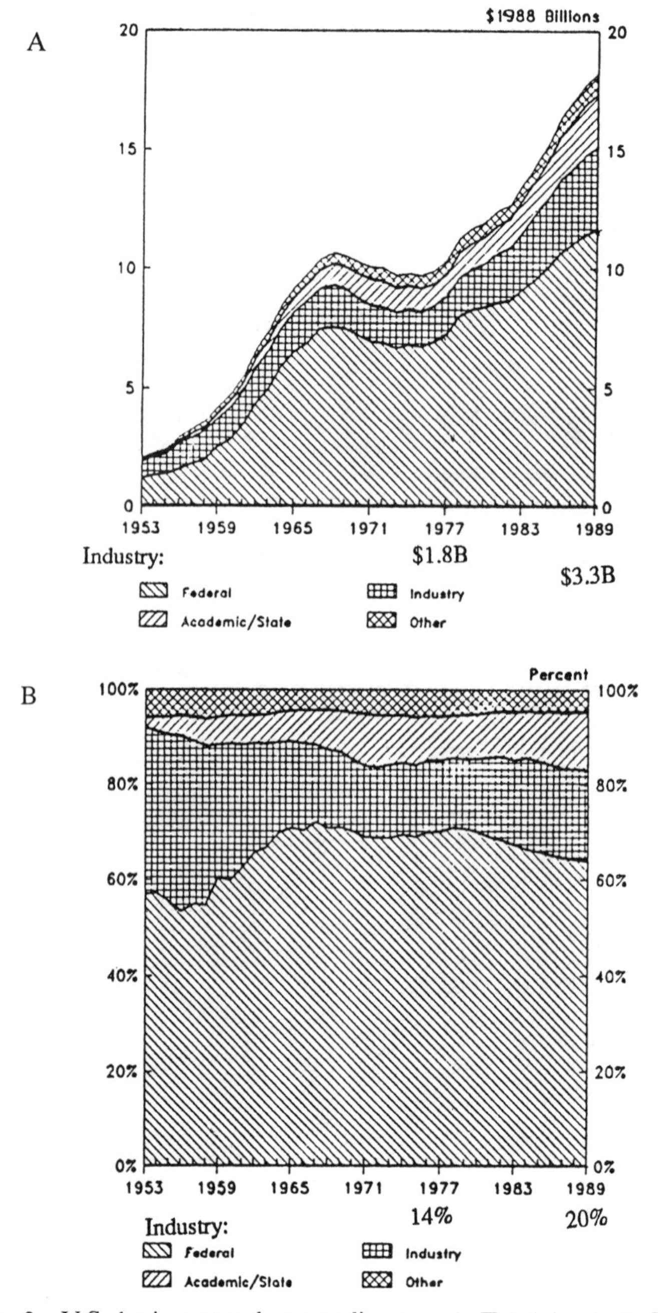

Figure 2. U.S. basic research expenditures. *A*, Total level; *B*, Distribution.

Source: "Science, Technology and the Academic Enterprise: States, Trends and Issues", The Government-University-Industry Research Roundtable, National Academy Press, Washington, DC, October, 1989.

ations of each partner. The expected outcome and the nature of the research planned should be articulated early in the discussions. These will serve as the basis for mutual agreement and will be critical in defining the type of partnership and its key parameters, including assignment of intellectual property rights, indirect costs, and mechanisms for interactions and measurement of results. An understanding or agreement can be derived on this basis to handle a range of expectations and research activities, as discussed below. Examples of types of university research which fall within general categories of the nature of research and the expected outcome, as shown in Figure 3, will serve as a framework for this discussion.

Major Program Types

Graduate fellowships are an excellent way for a university to enhance its capability for research and education without sacrificing any rights to freely publish results or to intellectual property and, if the research is fundamental and the sponsor is not expecting any competitive technology, without overhead costs. However, these arrangements generaly involve smaller funding levels for limited periods, and can sometimes be lacking in the sponsor's technical participation and its ability to provide specific applications.

Contract research, on the other hand, involves applied research, a commercial outlet for the results, and a framework for strong sponsor interaction, since the sponsor is expecting competitive technology to result. However, it may, in some cases, be too focused to take full advantage of the academic's expertise or to address strategic technical issues. While it solves a specific technical problem for the sponsor, it does not usually add to the technology base of the university.

A currently popular means of combining the favorable aspects of both of these above types is the University Center Program concept. A prominent example of this is the NSF Engineering Research Center Program, but many other smaller departmental interdisciplinary centers now exist. These programs usually fund applied research for generation of generic technology, which can attract multiple sponsorship, and in many cases is matched by federal or state dollars.

These centers have provided a means for industry to gain access to large interdepartmental resources, whose infrastructures and facilities are leveraged by public funds. The centers also provide a means of monitoring technology and access to students with multidisciplinary training. However, exclusive sponsorship usually requires additional arrangements and funding and the cost and bureaucracy associated with the center's overhead can be significant.

BP's Extramural Research (EMRA) Program

BP Research prticipates in all of these programs through its University Liaison Office at the Warrensville Research Center near Cleveland, Ohio. However, we also recognize that gaps still exist, part-

Table I

INDUSTRY-FUNDED UNIVERSITY RESEARCH

	1977	1988
As % of total industry research	2.7%	4%
As % of basic industry research	19%	27%

SOURCE: Reference 2.

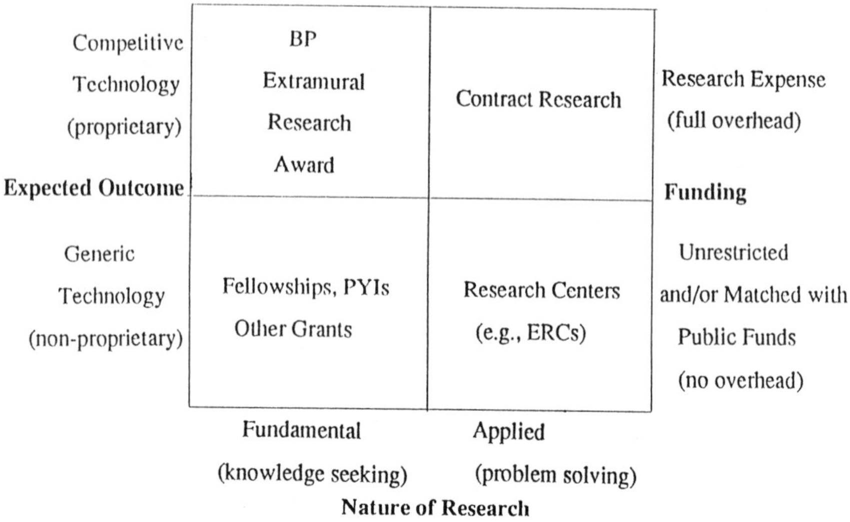

Expected Outcome	Fundamental (knowledge seeking)	Applied (problem solving)	Funding
Competitive Technology (proprietary)	BP Extramural Research Award	Contract Research	Research Expense (full overhead)
Generic Technology (non-proprietary)	Fellowships, PYIs Other Grants	Research Centers (e.g., ERCs)	Unrestricted and/or Matched with Public Funds (no overhead)

Nature of Research

Figure 3. Major types of University research partnerships and key parameters.

icularly is strategic areas where longer-term basic research is critical to the generation of competitive technology for one or more of BP's businesses. The EMRA program was designed to provide an integral component of a research base in areas of strategic and business relevance.

The EMRA program is a partnership designed to capture the benefits of academic freedom and commercial significance to BP. While BP chooses the areas where it needs to concentrate, the EMRA principal investigators actually write the proposals. In this task, the academics are given the freedom to pursue the approaches which will take advantage of their strengths in advancing the scientific understanding necessary to overcome critical technical barriers which limit the state of the art. The result is the development of strong contacts which academics working in areas of interest to BP which enhance not only the university funding base, but also the awareness of the relevance of its research to industry and of its researchers to practical, multidisciplinary problems. In turn, these strong academic links provide a key component of BP's fundamental research base and keep BP scientists at the "frontiers" of their discipline.

The EMRA proposals are developed by a procedure in which, after an internal review, selected prospective principal investigators are asked to provide their perspective on how fundamental research should be advanced to address a significant technical problem of strategic significance. Thus, the EMRA not only complements BP Research, but also advances the research goals of the principal investigator, and is written by the principal investigator, with general guidance on potential applications.

EMRA funding is intended to cover a three-year program and to involve at least one full-time researcher, in addition to the principal investigator's time. The program recognizes the university's need to publish and to share in the benefits resulting from EMRA's. It also establishes a mechanism for strong interaction of BP scientists with the EMRA team and for review of results.

The program has been in effect for about ten years in the U.K. It was initiated last year in the U.S. and has been received with enthusiasm by the academic community. At the time of this writing, 15 EMRA's have been placed in the U.S.

The EMRA concept provides another option for university-industry partnerships, where specific strategic technical problems and where an appropriate principal investigator and a relevant approach to advancing the understanding of the underlying phenomena can be identified. While BP in America has traditionally funded all the major types of university research programs discussed above, the EMRA program has already become an important part of BP's University Research portfolio (Figure 4).

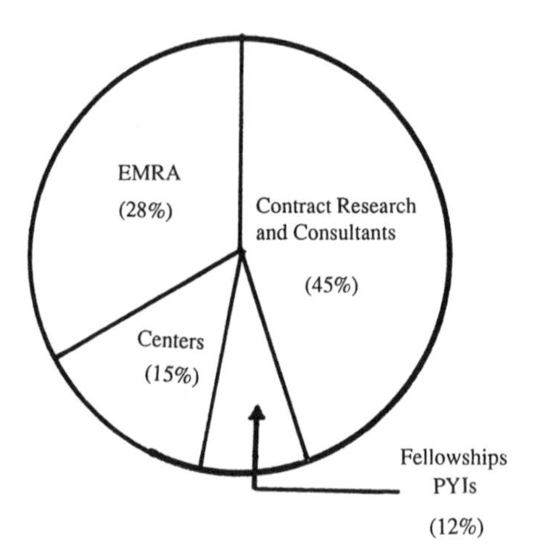

Figure 4. BP–University liaison, 1990 U.S. program distribution.

Prospects

While the spectrum of opinion on the topic of university-industry partnerships is broad, most participants and observers agree that there is value in these new collaborations. However, there is little concrete evidence that the enhanced activity over the past ten years has created real value, improved our competitiveness as a nation, or generated new jobs. While these are the ultimate measures of the partners' investments, it is still too early to apply the criteria to most programs.

As these programs mature, these questions will intensify and become more frequent. We should thus now be working to ensure that the appropriate data are collected which will comprise a definitive case.

For now, these partnerships can certainly be judged a success in terms of the increasing number of students able to deal with interdisciplinary problems, the enhanced technical interactions among academics and industrial scientists, the diffusing of issues surrounding the "incompatibility" of their respective cultures, and the renewed awareness of our interdependence in the process of innovation.

Literature Cited

1. *New Alliances and Partnerships in American Science and Engineering*; The Government–University–Industry Research Roundtable, National Academy Press: Washington, DC; 1986.

2. *Science, Technology and the Academic Enterprise: Status, Trends and Issues*; The Government–University–Industry Research Roundtable, National Academy Press: Washington, DC; October, 1989.

RECEIVED June 6, 1991

Chapter 10

Women and Minorities in Science and Engineering

AT&T Programs To Encourage Participation

F. T. Johnson, L. D. Loan, and D. W. McCall

AT&T Bell Laboratories, Murray Hill, NJ 07974

Public concern for the future vitality of United States' industry is receiving increasing emphasis in the press and on the agendas of professional society meetings (1-9). It is widely recognized that the availability of adequate numbers of talented and educated engineering and scientific people is a central issue and it seems clear that reliance on the traditional sources of technical personnel will not keep our enterprises afloat. While the popularity of engineering and the physical sciences as courses of study declines compared to medicine, business and law, white males (historically the group most attracted to science) will also decline to only 15% of the new entrants to the U.S. workforce in the period 1985-2000.

Women enter engineering and science in only modest numbers and their proportion has not increased in recent years. Minority participation in these fields is painfully low and shows no national tendency to increase. While immigrants have contributed greatly to past U.S. successes, it is not clear that the U.S. will be able to attract adequate numbers of the very best people as the technical competence of other nations increases. Thus, it is imperative that organizations in the U.S. put forth every effort to develop the human resources that will be necessary to maintain our technological future. In this report, several programs operated by AT&T Bell Laboratories will be described that bear on these issues. These programs are, and have been, extremely successful in exposing, attracting and sustaining young women and minorities in high technology pursuits.

The Bell Laboratories educational programs may be summarized as follows:

Schools:	Basic / eighth grade High Achiever / ninth grade High Step / tenth, eleventh, twelfth grade

0097–6156/92/0478–0095$06.00/0

College: Engineering Scholarship Program
Dual Degree Scholarship Program

Graduate: Graduate Research Program for Women
Cooperative Research Fellowship Program

Summer Employment: Summer Research Program for minorities and women.

Each of these programs will be discussed in the following sections. Although not relevant to the main thrust of this article, it should be noted that AT&T Bell Laboratories also offers an extensive collection of education programs for employees, both in-house courses and tuition reimbursement for attendance at local universities. Many employees have been supported as full-time students in a one-year Masters program. Employee programs will not be discussed further herein. Programs designed to assist in financing higher education for non-employees are also sponsored.

Schools Programs (10)

The Schools programs are characterized in Table I. The programs have been in operation for some 22 years and an estimated 90% of the participants enter college. The students, all women or minority, are drawn from the school systems of communities within easy commuting distance from Laboratories locations.

Table I. Characteristics of Schools Programs

Summer Science Prgms	Basic	High Achiever	High Step
Objective	*Create or Foster an Interest*	*Nurture Interest*	*Continued Motivation and Follow-up*
Academic Preformance	C	B	B+
Grade Completion	8th	9th	10th/11th/12th
Interest in Science or Math	X	X	X
Aptitude for Science or Math	X	X	X
Residence	Proximate to a Bell Laboratories Research Facility		
Self-Motivation	X	X	X
Recommendation from Head(s) of Science or Math Departments and Guidance Counselor(s) or Directors	X	X	X
Availability for Summer Science Programs	July	July	June through August
Parental Approval	X	X	X
Working Papers	---	---	X
Total participants '90	66	42	41

Transportation and lunches are provided. Laboratories locations at Holmdel, Murray Hill and Whippany, in New Jersey, and in Naperville, Illinois, participate.

The Basic program is intended to create or foster interest. The students, sixty-six in 1990, have completed the eighth grade and enter on recommendation of teachers and counselors. For two weeks in July mornings are devoted to technical projects under the direction of a host scientist or engineer. Laboratory tours and field trips take up the afternoons.

The High Achiever program is intended to nurture interest. The students have completed the ninth grade and a B-average is required in addition to school recommendations. Concentration on a single project under the direct supervision of a host scientist or engineer is pursued for two weeks in July. Forty-two students participated in the High Achiever program in 1990.

In the High Step program, the students work for ten weeks, as paid employees, on a closely supervised project. There were forty-one High Steppers in 1990. The purpose is to build motivation and enthusiasm for technical work. As a corollary, they function as useful assistants. The existence of groups of students with comparable levels of experience provides an opportunity to discuss their work socially and broadens the perspective of the individuals.

We believe that these school-level programs are effective mechanisms for instilling a knowledge of and interest in science and engineering. The technical hosts are challenged by the interactions and our evaluations have demonstrated that the experiences can be mutually rewarding. The success of these programs, and indeed all of the programs discussed herein, rests on an essential base of arrangements and logistical support provided by the Bell Laboratories Affirmative Action organization.

The students are generally well-behaved and eager to learn about science and engineering. They are unsophisticated in the early stages but become valued summer employees in the High Step sequence. Clearly, selection of projects of the correct level and abundant supervision are essential features. The students learn that technical work can be enjoyable and absorbing while difficult and demanding of discipline. The dedication of the mentors has an effect. The resulting influence on the students' attitude toward science and engineering is uniformly positive. For most of the participants there is no alternative exposure to technical work in a real-world setting.

AT&T Undergraduate Scholarship Programs

Undergraduate support programs are summarized in the Table II. Fifteen students enter the four year sequence, the Engineering Scholarship program, annually with full coverage of tuition, fees, books, room and board. Electrical Engineering, Computer Science, Mechanical Engineering and Systems Engineering are subject areas included, reflecting the bulk of AT&T future bachelor level employment (no specific employment commitment is implied, however). Attendance at colleges that offer strong curricula is required. Three students enter a five year sequence, the Dual Degree Scholarship Program each year to spend three years at Morris Brown, Spellman, Clark or Morehouse Universities and two years at Georgia

Table II

AT&T Undergraduate Scholarship Programs
Outstanding Minorities and Women

Provisions	Requirements	Disciplines	Selected per Year/Max
Engineering Scholarship Program (ESP)			
• Full tuition • Mandatory fees • Textbook allowance • Room & board • Summer employment • Housing arrangements (Summer) • Transportation (Summer) • Well-defined project • Mentor	High school seniors applying full time to college with strong curricula	EE, CS, Comp. Eng., ME, Syst. Eng.	15 / 60
Dual Degree Scholarship Program (DDSP)			
• Same as above • Receive dual degree	Attend Morris Brown Spellman, Clark or Morehouse for 3 years; Georgia Tech., Auburn, RIT or Boston U. for 2 years	BA in math or physics BS in EE, CS, Comp. Eng., ME, Syst. Eng.	3 / 15

Tech, Auburn, Rennsalaer or Boston University. The result is a B.A. in mathematics or physics and a B.S. in EE, CS, ME or Systems Engineering.

The undergraduate programs attract about one thousand applications each year, and they are consequently extremely competitive. Participants are largely black, hispanic and Native American but non-minority women (including API) are also included. In general these students do well in college and many successfully go on to graduate studies.

In addition to the academic year support mentioned above, AT&T undergraduate scholarship holders are offered summer employment of a form that is discussed below.

AT&T Graduate Scholarship Programs

The Cooperative Research Fellowship Program (CRFP) provides support for minority students engaged in doctoral studies in disciplines relevant to AT&T's research activities including chemistry, chemical engineering and materials science. Tuition, fees, books and an annual stipend of $11,500 (for the 1990-1991 academic year) are provided. In addition a summer work experience, mandatory for the first summer, in association with a Bell Laboratories mentor is part of the program. More will be said of this later.

The CRFP was initiated in the mid 1970's and grew gradually to its present size. About seventy individuals have received Ph.D.'s with CRFP support. This is a significant fraction of the total U.S. minority output in the fields covered. CRFP draws candidates from traditionally black colleges as well as institutions that have predominantly majority student bodies. Thus, to some extent there is a diversity of background. The awardees are all very promising academically. Even so, the students from the black colleges are apt to suffer "culture shock" on entry into an elite graduate university and the prior summer employment at Bell Laboratories helps prepare them and build confidence. About ten students enter the CRFP each year and it takes them about five or six years to obtain their Ph.D.'s. Throughout this period they maintain contact with their mentors who provide their main contact with Bell Laboratories.

The Graduate Research Program for Women (GRPW) offers fellowships with the same support provided under the CRFP (see Table III). Four students enter GRPW each year with fellowships and they also take five or six years to complete

Table III

**AT&T Graduate Scholarship Programs
Outstanding Minorities and Women**

Provisions	Requirements	Disciplines	Selected Per Year
GRPW & CRFP Fellowships			
• Tuition and fees • Textbook allowance • $11,100 annual stipend • Summer employment • Housing arrangements (Summer) • Transportation (Summer) • Well-defined project • Mentor • Conference travel	Seniors applying to graduate school for Ph.D.	Chem., Chem. Eng., Comm. Sci., CS/Eng., EE, Info. Sci., Matl. Sci., Math, ME, OR, Physics, Stat.	GRPW: 4 fellowships CRFP: 9-12 fellowships
GRPW Grants			
• $1,500 annual award • Summer employment • Housing arrangements (Summer) • Transportation (Summer) • Well-defined project • Mentor • Conference travel			GRPW: 6 grants

the Ph.D. requirements. The women in GRPW tend to be non-minority (white and Asian) as minority candidates gravitate to CRFP where there are more fellowships. In addition to the fellowships, GRPW offers six grants each year ($1,500/year). GRPW grant holders obtain principal support from other sources (e.g. NSF fellowships, teaching assistanceships or other scholarships). Both grant and fellowship awardees take part in summer employment (mandatory for the summer prior to graduate school) and have the benefit of mentors who maintain contact during their doctoral studies.

The applicants for GRPW are characterized by unusually strong undergraduate records and many come from the most competitive universities. In view of these impressive credentials one might expect adjustment to a leading graduate department to be routine, but it is not. Science and engineering departments tend to be populated primarily by men and few women occupy tenured positions in leading departments, although some progress has been made in this regard in recent years. Women tend to feel unwelcome. There is some feeling that male professors may be reluctant to give their best ideas to women as thesis topics. For various reasons, women are not always comfortable in leading engineering and science graduate departments and the mentors can often help them settle in effectively.

The GRPW was begun in 1975 and about sixty women have been awarded Ph.D.'s under the program. Many of the graduates have chosen to pursue academic careers and a few are now employed by AT&T.

Summer Employment

The Summer season at Bell Laboratories is exciting owing to the presence of a substantial number of bright young students. In addition to the fellowship holders, a Summer Research Program (SRP) designed for outstanding minorities and women brings 60 – 100 individuals who have typically just completed their junior year of college (see Table IV). The SRP covers essentially the same fields as

Table IV

AT&T Summer Employment Programs
Outstanding Minorities and Women

Summer Research Program (SRP)

Provisions	Requirements	Disciplines	Selected per Year/Max
• Summer employment • Housing arrangements (Summer) • Transportation (Summer) • Well-defined Project • Mentor	College juniors (and seniors) not graduating by May	Cer. Eng., Chem., CS/Eng., EE, ME, OR, Chem. Eng., Physics, Math, Info. Sci., Comm. Sci., Stat., Matl. Sci.	60-100

GRPW and CRFP. Several discipline-specific committees composed of Members of Technical Staff select SRP participants based on academic achievement, personal statements, relevance of interests, and recommendations. Each student is carefully matched to a pre-arranged research project and mentor. During the summer the visitors are housed in Rutgers University dormitories (about 15 miles away) and buses are provided to take them to and from work. Lectures are offered covering laboratory safety, orientation in regard to Bell Laboratories generally and specific technical seminars. Social outings are also arranged.

Individual projects are designed such that they can be completed in the ten-week period. This gives the student enhanced satisfaction and it often results in publication of a note or paper. The SRP students finish their summer work by giving seminars to the assembled group. This event is the cause of considerable tension before and a great feeling of accomplishment after (see Table V).

Table V
SRP / GRPW / CRFP Calendar of Activities

October – December	Special Programs recruiters contact and visit target colleges.
November – December	Mentors and project proposals for next wave of applicants are evaluated by program and discipline committees.
January	Applications are received and distributed to appropriate program and discipline committees.
February – March	Applications are studied and selections made for participation by program and discipline committees. Offers are made to SRP selectees by discipline committee members. Candidates for CRFP and GRPW are interviewed at the laboratories and final offers made by program committees.
April – May	Staff workers make logistical arrangements for transportation, housing, employment and other necessary factors.
June – August	Students in residence, complete projects, participate in orientation, broadening seminars, social outings of groups, etc.
August	Final seminars, return/embark on school year.

Conclusions

The foregoing description of AT&T programs for the enhancement of minorities and women in physical science and engineering careers is brief and may not convey the enthusiasm that is apparent among individual employees who manage and implement the programs. Indeed, the initiation and development of these programs owes much to the initiative and dedication of individual members of staff at Bell Laboratories. These people initially "sold" the programs to management and continue to work hard and long hours to make them successful. The credit for the existing programs is theirs. The technical staff and the administrative staff work closely and in harmony. Long hours are spent poring over the applications and debating the prospects of candidates for future success. The emphasis is on what the programs can do for the student. Useful work is a bonus. AT&T employees learn from the experience and the laboratory atmosphere is improved by the presence of more diverse viewpoints. The programs are important and the nation needs further examples in kind.

References

1. Milbank, D., "Shortage of Scientists Approaches a Crisis," Wall Street Journal, September 17, 1990.
2. Holden, C., "Wanted: 675,000 Future Scientists and Engineers," Science 244, 1536 (1989).
3. Vetter, B., "Women in Science, Progress and Problems," AAS Symposium, January 15, 1990.
4. Vetter, B., "American Minorities in Science and Engineering," Commission on Professionals in Science and Technology, (1500 Massachusetts Avenue, N.W., Suite 831, Washington, D.C. 20005).
5. "Education and Employment of Engineers, A Research Agenda for the 1990's," Steering Committee on Human Resource Issues in Engineering, National Research Council, National Academy Press, Washington, D.C. 1989.
6. Pool, R., "Who will Do Science in the 1990's," Science 248, 433 (1990).
7. Wycliff, D., "Science Careers are Attracting Few Blacks," New York Times, June 8, 1990.
8. Atkinson, R. C., "Supply and Demand for Scientists and Engineers," Science 248, 425 (1990).
9. Bloch, E., "Education and Human Resources at the National Science Foundation," Science 249, 839 (1990).
10. In addition to the schools programs covered herein, AT&T Bell Laboratories at Murray Hill, NJ offers a series of four or five Saturday seminars each year: "The World of Science Seminars." This series is intended for high school students and their teachers from surrounding counties. Typically, 200-400 attendees are present. Each lecture/demonstration is given by a scientist at AT&T Bell Labs on a current research topic. The series began in 1984.

APPENDIX: Case Histories

CRFP Composite Experience: Bill attended a small, predominantly black, undergraduate school and first heard of AT&T Bell Laboratories from a Fellow student who had spent a summer in Murray Hill on the SRP program. From him he heard of the CRFP program which provided graduate school support and when the Special Programs Recruiter arrived from Bell Labs, he decided to apply. He took the GRE exam late in the year and sent off application forms in mid-December.

In early March he was invited to Murray Hill for personal interviews and was told by his host to consider each of the interviewers as potential mentors. He further explained that one special part of the program was to foster a relationship between each Fellow and a staff member who would be the Bell Labs contact through graduate school. During the day Bill met a number of people and was particularly impressed by Joe, a young member of staff, who had worked in an area of research Bill found very attractive. Bill had applied to several graduate schools before his interview but was very interested in the help he received during the day on how to choose among them.

In early April Bill was very pleased to hear from the committee members that he had been awarded a fellowship and they agreed that Joe would be his mentor.

After a summer working with Joe at Murray Hill, Bill entered graduate school. During the course work he and Joe talked every month or two - more frequently on one or two occasions when he had a conflict with one of the professors or some worries about his progress. In each case Joe was ready to help and smooth out some rough patches. When the time came to choose a thesis advisor, Joe's input on relative reputations, areas of research, and on current "hot" areas of work was very helpful, and Bill finally made his choice.

On his way back to school after the first year Bill spent a couple of days at Bell Labs catching up on Joe's work and renewing acquaintances. He could have spent a second summer working with Joe, but preferred to get a jump start on his research. Starting up his project was not entirely straightforward and having Joe to talk to was helpful. After several months of very hard work, Bill was happy to find results coming quite nicely. So much so that his advisor suggested he give a poster presentation at an American Chemical Society (ACS) meeting. Bill talked this over with Joe and discovered that Bell Labs was prepared to pay for his attendance, Joe was one of the visitors to his poster.

Bill was surprised at how quickly his five years in graduate school passed. Towards the end when he started out on job interviews, he again used Joe as a partner in discussing job opportunities. He was approached by several recruiters from various companies including one from Bell Labs. He knew this last one reasonably well since she had visited him each year during her visit to campus. After a number of interviews and company visits, Bill finally decided on a job at Bell Labs, maybe because he knew it best. He came to the choice on his own and he was glad there had been no pressure put upon him to make it.

It is now several years since Bill started work. He has worked on a variety of projects and with many different people, two were also CRFP Fellows.

GRPW Experience: Joan heard about the Bell Labs GRPW when she was introduced to a special recruiter in her department chairman's office in October. Joan was a senior at an elite eastern undergraduate college and her transcript was straight A's. Her GRE's were in the high 700's for both verbal and quantitative and she planned doctoral studies in chemistry. She later applied for a GRPW grant because she was confident of obtaining an NSF or other industrial fellowship for her primary support. She also applied for admission at Harvard, Cornell, Illinois, Cal at Berkeley and Stanford. During February, Joan received a telephone call in which one of the Bell Labs chemists on the Materials Division committee, Mary, invited her to visit Murray Hill, New Jersey, for interviews as a finalist in the GRPW competition. In early March Joan took the train to Metropark, a suburban station on the Amtrak mainline, and was met by her host, Mary. The interviews began at dinner, filled the next day breakfast through dinner, and went on for half the following day. On that afternoon Joan returned to school by train, exhausted but stimulated. Two weeks later she received a phone call from Mary and learned that she had been selected as a GRPW winner. Mary told her that it was proposed that her mentor would be Tim. Joan had spent two hours with Tim during her interviews and she was clearly interested in his research. She was excited about spending the summer doing research at Murray Hill. On June 1, Joan again went to Metropark where she was met by Harold of the staff organization. Harold took Joan and two other students to the Rutgers University dormitory that was to be their summer home. The dormitory was crowded with over 200 'strangers' from many parts of the country who had converged for the SRP, GRPW, CRFP and other programs. Students were assigned in congenial groups in four-student apartments consisting of two bedrooms, a living room and kitchen. The next day buses drove the students to Murray Hill (Whippany or Holmdel). Joan spent the morning getting a physical exam and listening to orientation lectures. Tim and Mary took her to lunch and then to her laboratory. A desk and telephone had been arranged. Following introductions to other chemists in the area, Joan took the bus back to the dormitory with an armload of books and reprints.

The next morning research began in earnest. Joan's project was well-defined and Tim was always available to help. The weeks passed quickly and Joan began to understand the project and why it was important. Gradually, she began to worry about August and the necessary seminar. Social life at the dormitory was active and twice during the summer there were theater bus trips to New York City. Joan, with Tim's help, began to develop her seminar as July wore on. She discovered that there was a high level of anxiety in regard to the seminars among the other students, particularly the SRP's who were a year behind academically.

Two days of seminars were scheduled to accommodate the Materials Division students. The audience, about eighty, consisted of the students and mentors, along with other interested Bell Labs scientists. All went well and Joan emerged with new confidence and enthusiasm for chemistry. The next day, her last at Murray Hill that summer, Joan and Tim finished a note to the Journal of the American Chemical Society and put it in the mail. Joan then returned to her home in Pennsylvania to prepare for her entrance into Stanford.

RECEIVED February 15, 1991

Chapter 11

Keys to Successful Partnerships

Alan L. McClelland[1]

Education and Human Resources Directorate, National Science Foundation, Washington, DC 20550

As in most partnerships, business/education partnerships must be based on certain principles to succeed. There are four that seem crucial for improving science education through the cooperative efforts of teachers with scientists and engineers from industry: (1) good intentions alone are insufficient, (2) the partnership program must be based on an accurate understanding of what the real problems in science education are, (3) each partner must respect and appreciate the expertise of the other, and (4) science must be taught in a way which truly engages the students' interest.

In 1987, the National Science Foundation initiated a program of "Private Sector Partnerships to Improve Science and Mathematics Education" to draw on the intellectual resources of the private sector. Since two-thirds of all scientists and engineers in the United States are employed in private industry, where they carry out three-fourths of all U. S. research and development, the private sector constitutes a major intellectual resource on which the educational world should be drawing. With over $18 million now committed to more than 70 projects, the staff of the Education and Human Resources Directorate of the National Science Foundation has had ample opportunity to observe how successful partnerships between educational institutions and private sector firms and organizations can be. Likewise, we have also been able to identify factors which can keep such partnerships from accomplishing as much as anticipated.

The majority of the partnership projects funded through this program have been at the elementary and secondary level, though a few have been at the postsecondary level. On the educational side, they have involved

[1]Current address: 22 Guyencourt Road, Wilmington, DE 19807

0097–6156/92/0478–0105$06.00/0

K-12 schools, 2- and 4-year colleges and universities, museums and teacher's organizations. In most cases the organization actually submitting the proposal to the NSF has been an educational institution; very few have come from for-profit industrial or business firms, though all have involved such private sector organizations as partners. In fact, there is no reason, legal or practical, which should prevent a company from being the applying organization, and we hope to see more such applications in the future.

Most of the projects have been quite successful and through them the program has clearly demonstrated how valuable inputs from outside the educational world can be for science, technology and mathematics education. Much, perhaps most, of the current content of education at all levels has been established by the educational world itself, particularly the college/university segment. This arises both from the way future teachers are educated and from the role of university discipline departments in recommending curriculum content. There are many positive aspects to this, but since it is basically an inbred relationship, it carries with it the dangers, as well as the strengths, inherent in inbreeding. The influence of technical people outside the educational world can be a valuable balancing force in helping establish educational directions, particularly since by sheer numbers they constitute the mainstream of technical activity.

Out of these experiences with a substantial number of partnership projects, I have drawn some personal conclusions about what works and what doesn't. With an appreciative nod to Ben Franklin and his Poor Richard's Almanac, I'll try to group my conclusions under four familiar adages or aphorisms or proverbs. They may seem simplistic, but I truly believe reflection on them by anyone, whether from the educational sector or the private sector, interested in developing a partnership can help avoid some serious pitfalls.

The Road to Hell is Paved With Good Intentions. The conviction that our present patterns of science and mathematics education are not meeting the needs of the nation today is widespread, both inside and outside the educational world. Accordingly, many well-intentioned people feel moved to help solve the problem -- very commendable, but insufficient in itself. "The Road to Hell ----".

One way to pave that road is by failing to recognize how much hard work has to accompany the good intentions. Someone is going to have to put in a lot of time, effort, and, yes, often money -- big problems are not solved with small efforts. We see time and again in the successful partnerships the dedicated effort of one or two or three people. Some chemists who have done this in our projects: Marwin and Nan Kemp and Eric Bandurski at Amoco in Tulsa, who have developed a strong elementary science program utilizing many volunteers; Nina Klein at Montana College of Mineral Science & Technology, who has drawn on the mining industry in southwest Montana to relate science teaching to the most important local industry; Joseph Bieron at Canisius College in Buffalo who has worked with local industry to make instrumental analytical equipment and experiments readily available to high school teachers. Each partnerships will require some prime movers who will back up their good intentions with hard work.

Another way to build that "Road to Hell" is to aim in the wrong direction -- tackle a "problem" which really isn't a problem, or at least not a very important one. Which brings me to my next maxim.

The Plural of 'Anecdote' is Not 'Data'. I'm indebted to Iris Weiss of Horizon Research, Inc., for this one. It's amazing -- and depressing -- how many of us outside the educational world will pontificate on what's wrong based on a few anecdotes -- "my son _____", "my neighbor's youngster ___", "a teacher I know said _____". It's a trap we technical people ought to know enough to avoid --- overgeneralizing on the basis of one or a very few experimental observations.

Let me give you one example. A current widely repeated lament -- "college enrollments in the technical fields have been declining in recent years". Look at Figure 1 and Table I. The message: enrollment trends vary greatly by discipline. Yes, students are losing interest in majoring in chemistry and physics (B.S. degrees in both peaked in 1970). Biology had a huge rise until the mid-seventies and has dropped since (I argue because of a shortage of jobs -- college kids aren't so dumb!). Engineering, though, has clearly been on an up-trend since the mid-fifties, with a very sharp rise in recent years. True, it has been dropping off in the last three or four years, but I predict that will be a temporary drop if job demand picks up.

Another myth these data lay to rest is that young people are afraid of math-heavy subjects -- that's how most chemists and physicists explain the decline in their fields vis-a-vis the biological sciences. But engineering -- a very math-intensive area -- is the fastest growing of all the technical fields. Incidentally, the fastest growing of all fields, and now the number one college major, is business, which points out one input industry technical people in a partnership could document and underscore -- the best route into most technically-based companies is through science or engineering (e.g., in Du Pont, my former employer, 80% of all the college graduates have degrees in science or engineering).

The key point here is not to argue the details of the above, but to urge anyone going into a partnership to get a strong factual base for understanding the other side. Educators need to make an effort to get some understanding of industrial organization -- who does what, who makes what kinds of decision -- while industry people must get some real knowledge of the schools they intend to work with. Just because you once went to school doesn't mean you know all about education! For an educational data base I recommend the Digest of Educational Statistics, put out almost every year since 1962 by the National Center for Education

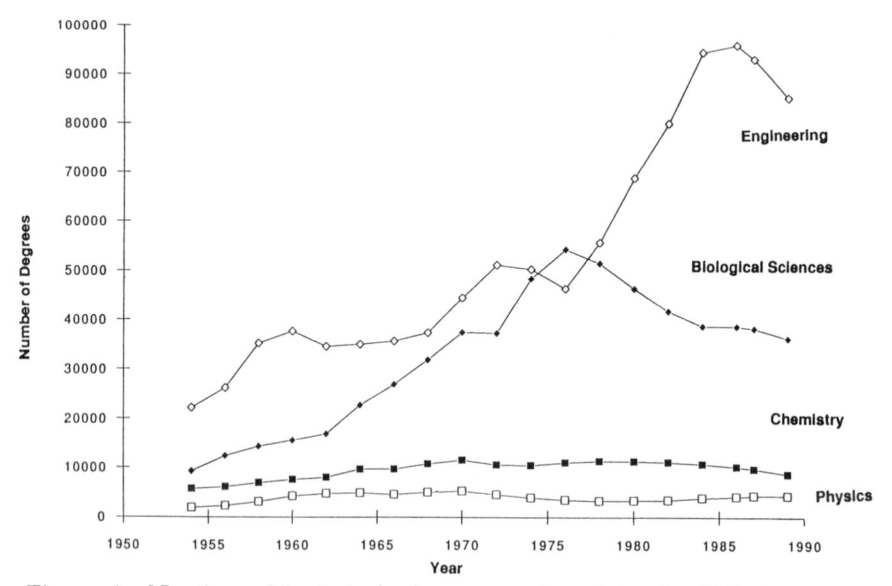

Figure 1. Number of bachelor's degrees conferred in the U.S. between 1950–1990. Enrollment trends vary greatly by discipline.

TABLE 1. BACHELORS DEGREES: UNITED STATES

Year	Chemistry	Physics	Biological Sciences	All Engineering	All Fields
1954	5752	1949	9279	22227	290825
1956	6141	2329	12423	26219	308812
1958	6982	3179	14308	35191	362554
1960	7569	4322	15576	37679	392440
1962	8047	4808	16915	34551	417846
1964	9660	4946	22723	35013	498654
1966	9687	4601	26916	35615	519804
1968	10783	5038	31826	37368	632289
1970	11519	5320	37389	44479	792316
1972	10590	4634	37293	51164	887273
1974	10438	3952	48340	50286	945776
1976	11022	3544	54275	46331	925746
1978	11315	3330	51502	55654	921204
1980	11232	3396	46370	68893	929417
1982	11062	3472	41639	80005	952998
1984	10704	3907	38640	94444	974309
1986	10116	4180	38524	95953	987823
1987	9661	4330	38114	93074	991339
1988	9025	4097	36761	88791	993362
1989	8654	4339	36079	85273	1017667
Increase 1989 vs. 1954:					
	2902	2390	26800	63046	726842
	50%	123%	289%	284%	250%

SOURCE: National Center for Education Statistics, U. S. Department of Education. (See various annual issues of *Digest of Education Statistics*, available from U. S. Government Printing Office.)

Statistics, a unit of the U. S. Department of Education, and available from the U. S. Government Printing Office. The data above came primarily from the 1989 edition, a 462 page treasure trove. It makes fascinating reading!

Some of our best partnerships are based on helping the two sides to truly understand the needs and characteristics of the other. One kind of example: programs which place teachers in summer jobs or internships in industry, allowing them to get a first hand experience in how industry uses science and mathematics. But a clear message has come out of those projects: it's not easy to translate that industrial experience into classroom-useful material. Thus while the teachers are learning what goes on in industry, the industrial scientists, engineers, and technicians have to learn enough about those teachers' environments to develop useful assignments and follow-up activities for the teachers. Examples of programs which have been particularly successful at this are the Industrial Initiatives for Science and Mathematics Education operated in the San Francisco area through the Lawrence Hall of Science, a program in Michigan operated out of Grand Valley State College, and the New Jersey Business Industry Science Education Consortium, a partnership of numerous industries and schools headquartered at Stevens Institute.

It Takes Two to Tango. Like any successful marriage, a partnership must have an essentially equal commitment by each partner. Too often an enthusiastic person on one side will talk people and institutions on the other side into forming an unbalanced partnership. Equally important is true agreement on the goals; not only do both partners need to want to dance, they both have to want to tango -- or square dance, or waltz, or whatever is the program of the evening. Thus, early on, clear goals and program details need to be formulated so the right dancers are recruited. Examples of targeted programs which successfully recruited participants committed to the partnership activities include: development of coordinated curricula on environmental issues related to industrial and municipal waste disposal in grades 4-12 of the Haywood County, NC, schools; encouragement and subsidization of the introduction of modern molecular biology experiments into high schools by Edvotek, Inc., and the National Association of Biology Teachers; recruitment by the Saturday Academy of the Oregon Graduate Center of industrial sponsors for invention projects by elementary and middle school students.

Frustrating as it may be, it is crucial for the person and/or organization which wants to initiate a partnership to keep searching until the right partner is found. Even two good dancers may not dance well together, but when you find the right partner you usually know it.

"I Hear and I Forget; I See and I Remember; I Do and I Understand". This allegedly ancient Chinese proverb carries a message most technical people should find appealing: "hands-on", experimental activities have the most impact. Industry is full of so many fascinating applications of science and mathematics that development of activities that students can actually do is a very appropriate partnership activity. This may involve industrial people in developing hands-on activities for teachers, who in turn can take these activities to their students, or it can involve working directly with students. An excellent example of the former is a project in the Northmont City School District in Ohio where local industries, including Mead Imaging, EG&G Mound Laboratories, and Dayton Power & Light, worked with elementary teachers to develop an elementary science laboratory for the district and modules using it for all the elementary grades. The latter is exemplified by a project developed through Lesley College in Boston to involve middle school girls in science activities led by industry and museum scientists, particularly women scientists, to encourage them to take science courses through high school and consider possible science careers.

A key caveat here: it is teachers and others (e.g., science museum staffers) who work directly with students who should be the final judges of what's right for any particular group of students. True partnership is necessary here: industry people suggesting possible activities, but working closely with their educational partners to pick out those that have the most promise of being effective. A prime example comes from a National Science Teachers Association project funded by the NSF prior to the initiation of the Private Sector Partnerships program. In the summer of 1984, sixteen high school chemistry teachers participated in summer workshops at three industrial firms (ARCO Chemical, Du Pont, and Shell Development) on polymer chemistry; subsequently they wrote a 244 page book, *Polymer Chemistry, A Teaching Package for Pre-College Teachers*, based on their workshop experiences, which was published by the NSTA and has been widely used in schools across the country.

Partnerships involving educators and those utilizing science and mathematics in the non-educational world can make a major contribution to insuring that science, mathematics, and technology education truly prepares students to leave the educational world well-equipped for what they will encounter there, rather than just preparing them for the next level of school or college. To succeed, however, such partnerships must be based on more than just good intentions -- the partners have to be prepared to really work toward goals based on a true understanding of well-defined needs and problems, not superficial judgements based on anecdotal evidence. As in all cooperative activities, the partners must be compatible and equally committed to those goals. And finally, it should never be forgotten that science is best learned through the hands in addition to the eyes and ears.

RECEIVED May 20, 1991

Chapter 12

Science Demonstrations

Melanie J. Cravey

Department of Marine Sciences, Texas A&M University at Galveston, Galveston, TX 77553

Nalco Chemical Company and Texas A&M University at Galveston have cooperated on the assembly of science demonstrations to present at elementary schools. The emphasis is on the three states of matter, changes of state and the three dimensional nature of molecules. While geared for the third grade level, these demos are appreciated by students and their teachers even into high school. Why this age level was chosen and how the program is managed are discussed. An outline of the presentation is provided and some suggestions for other ways to bring science to the community. Appendix I is a list of some additional sources of demonstration ideas.

The joy of discovery is an intense activity for children. They want to touch and try all manner of new thins. Thus, it is an ideal time to introduce them to science and its endless wonders. Unfortunately, the typical classroom presentation contains dry facts to memorize and neglects the discovery approach. Somehow the fun aspects of science must be brought back to the classroom. But we aren't doing anyone any favors if we don't tie the excitement of "hands-on" science with the fundamentals of "how and why?".

Hopefully by now all the members of the American Chemical Society and others in the scientific community have gotten the word that there are not enough science and engineering students coming down the educational pike to fill the need for the next 15 years. **WE** must convince more young people to make science or engineering a career goal, and (ideally) simultaneously affect the attitudes of students, their parents, teachers and community favorably toward science. But what age students should we target?

0097–6156/92/0478–0111$06.00/0

How do we get started? What do we say and do? How do we find other people to help? In this paper the above questions are addressed, our program is outlined and other sources of materials and ideas are discussed.

Texas A&M University at Galveston and Nalco Chemical Company in Sugar Land cooperatively developed a program especially for third graders. (It is suitable, however, for presentation to many ages simply by adding experiments and increasing the technical level of explanations for the various phenomena shown).

Choosing a Target Audience

In order to choose a target audience the program goals must be decided. Two obvious goals are: (1) to persuade more students to choose science or engineering as a career and (2) to favorably impress the public with science as beneficial to both individuals and society. It seems quite likely that if we make the second goal primary (make science popular), the other goal (become a scientist) will naturally result (more people will want to be associated with it). Realistically one might consider the source of funds and add a capitalistic goal, ie. sponsoring organizations want to enhance their image and recruit future employees or students. But if in fact science becomes more popular then the sponsors (groups of scientists doing science) will also. Thus the original goals are still sufficient. The added benefit may be name recognition for small schools or companies, or those which do not sell directly to the public.

To make science and engineering popular requires two components: persistence and credibility. There are two reasons for persistence. First, the news media tends to focus upon negative events and keep that in the public mind. To counteract that we must continually bring the benefits of science into public view. Secondly, studies have shown that students make decisions in junior high that will ultimately determine their preparedness for science or engineering. To be successful in college they need three years of science and math in high school, which assumes a strong junior high background. Thus targeting a program only at high school students is probably too late for some.

The reasons to start in elementary school are primarily psychological. Kids are curious, have no preconceived ideas about the subject and their attitudes are still quite flexible. Also most teachers are happy to include special programs and their schedules are easier to interrupt than in higher grades. Simple, safe demonstrations are available which easily fit into the elementary science curriculum.

Credibility is crucial to the improvement of the public view of science. Honesty about hazards and a sincere presentation force the demonstrators to make

Feb. 7, 1991

Dear Docter Cravey,

We appreciate the stickers! Thank you for showing me chemistry and Science very much! It was very interesting. I liked how you made slime! And how you poped the potato out of the tube! I didn't like science until you explained it to me!

Your friend,

Aaron F.

reasonable explanations of whatever phenomena are shown at a level appropriate to the audience. It has been the experience of many demonstrators that an unfortunately large segment of adult lay people are enlightened and entertained by programs we think more suitable for young children.

The best way of influencing the most kids for the amount of time invested may be teacher in-service workshops. Many elementary teachers (and even higher grades) do not have a good grasp of scientific principles, and through such workshops their knowledge base can be improved upon while providing them with the tools to take back to the classroom. This is a good thing to do for the community and the school district, but there may be a need to keep supplying materials to the teachers every year to keep the program viable. On the other hand, the students really like to see a real scientist and ask questions. The impact will perhaps be more memorable if you come to the school than if you just teach the teachers.

Based on criteria such as these and input from real kids we decided to go for the elementary school demonstrations. However, we have done these same demonstrations at malls for National Chemistry Week, for high school and college students, teachers, administrators and even colleagues. The real charm of chemistry and physics works on everyone!

Choosing the Demonstrations

There are an immense number of demonstrations available. After all, chemistry was taught long ago primarily in lecture halls equipped with demonstration tables. Although the sciences have moved into more pedantic presentations of material, there have long been people active in keeping the art of successful demonstration alive and well. Ideally one does not want to "reinvent the wheel", so finding the available material is an excellent starting point. For the convenience of the reader a list of sources has been compiled as Appendix I at the end of this chapter. Hopefully each person who becomes active will contribute ideas and innovations into the community pool.

An integral part of choosing demonstrations is deciding exactly what you want to accomplish. Consider these interesting notes.

- Our society has a fundamental difficulty with observation and description as applied to problem formulation and solution.
- Most people do not realize that molecules are three-dimensional.
- Many people have no feeling for probability or equilibrium (both of which have many uses outside of pure science).

Perhaps one reason science is not more popular is that the processes scientists use in their work are not understood.

Taking these notes into account might lead you to choose demos which allow you to address at least some of these issues.

Find out about the curriculum of the target students. If at all feasible, plan demonstrations which will reinforce the material they are studying, preferably building thereon and taking them farther than the teachers can due to their limited resources or experience. Such a philosophy will facilitate your acceptance by both teachers and administrators, as well as providing the students a logical bridge to new concepts. For example, third graders study the states of matter. For this target audience the teachers were told our "lesson plan" was to demonstrate that air molecules have mass and can do work; the relationships between the states of matter; and that molecules are three dimensional.

Choose experiments which are not too complicated for the target audience. They need not be expensive and should be easy to explain to other demonstrators or teachers so the work can be carried on by as many volunteers as possible. Select enough demos to make the show last about forty-five minutes to an hour. Try the experiments out ahead of time, and be prepared to drop some that are difficult or unreliable. Replace them with others as you acquire experience and as new materials become available.

The Demonstrations We Use

Our experiments are organized such that we go through our "lesson plan" (air has mass and can do work, the relationship among three states of matter, and molecules are three dimensional) in an orderly fashion. Commercial applications of all the science demonstrated are discussed. Remember we want science to be viewed as useful, important and understandable. The selections continually evolve as we modify or exchange, but what follows is a representative list of experiments and a few words about the points each is to address.

1. Introduction with Invisible Ink Sign[1,2] - science is fun and important. It meets our basic needs for food, clothing, shelter, and even life processes depend upon it. Curiosity and observation are necessary tools for a scientist.
2. Clock Reaction[3] - this is set up early so they can watch it go through a number of cycles. We picked one with four color changes, and later discuss the concepts of redox and equilibrium (or balance).
3. Blue Bottle[4] - again early set up gives plenty of observation time. Also relates to redox and equilibrium, and can be compared to how blood works to carry oxygen.
4. Egg in the Flask[5,6] - air is in the flask (it is

Dear Dr. Cravey,

Thank you for coming to Westwood and doing the chemistry experiments. I thought the dry ice skipping the liquid state was neat. I also liked how the little crystal soaked up all that water and kept getting bigger and bigger. Those were the best experiments I've ever seen.

Sincerely,

Stephen Hebert

not empty), which can be expanded with heat so that some leaves the flask. When the egg blocks the opening and the flask is cooled, the contracting air creates a partial vacuum which pulls the egg into the flask. Reversing the process pushes it out again. Thus work is being done by the air outside the flask acting to relieve the vacuum in the flask.

5. Potato Gun - air is again doing work, this time through compression (like a pneumatic tube at the bank).
6. Bottles and Balloons - a plastic bottle modified to have two mouths and two balloons can be used to show relationships between pressure and volume. (Wonderful even for freshman chemistry in college).
7. Dancing Raisins[7] - also shows air doing work, lifting the raisins. Compare to "floaties" for kids. (Doesn't work with sugarless drinks).
8. Melting Ice and Sublimation - solid water changing to its liquid state and solid carbon dioxide changing to the gaseous state. Naturally there are plenty of fun things to do with dry ice, but don't let anyone get burned.
9. Disappearing Coffee Cup[8] - another use of air (as an insulator) and the process of dissolving. Other things made mostly of air are shaving and whipping creams.
10. Crystallization of Sodium Thiosulfate[9] - an impressively rapid change of state, but can be unreliable as it tends to crystallize before you get to the demo site.
11. Superabsorbent[10] - compare the process of dissolving a solid in a liquid (e.g. sugar in water) to the superabsorbent (liquid "dissolves" in the solid).
12. Diaper[10] - practical use of superabsorbent and good source of humor.
13. Water Thickener - compare to jello. Can be very impressive when properly performed.
14. Slime[11] - everyone's favorite and very easy. Even young audiences get the idea that there are attractive forces between molecules after these two demos.
15. Beads and Nylon - use beads to represent small molecules and long polymers[12], then make nylon[13] and compare to the beads. (I plan to substitute latex for the nylon).
16. Electrochemistry - in a small group you can show the potential in a potato or lemon battery using a volt meter, but it may be less useful in a larger group.
17. Needle in a Balloon[14] - compare the pile of plastic beads in #15 to the polymer of the

balloon. There are spaces between the molecules, and the polymer can be gently pushed aside (or few enough bonds are broken) to allow the entry of the needle.

18. Magic Cups - this is the only time the word magic is used. Hopefully by this time they know there is a logical explanation. The showmanship of properly going through the cups is fun for all.

Occasionally we add other experiments, especially for older audiences. For example a silver fractal[15] can be grown on site and left for them, as a substitute for the electrochemistry. It works well for freshmen chemistry students also, if discussions about the precipitates and equilibria are included. A fun physics demo to add is a bicycle wheel if you have one (it needs to have an axle with handles), to show momentum and torque. Some demonstrators like to take along molecular models for use as they discuss the three dimensional nature of molecules.

Remember that teachers, parents and principals will see the demonstrations. We therefore studiously avoid the words "magic" and "wizard". We have very little smoke, few pops and no flash. The changes of state, color and texture (they are allowed to feel the superabsorbent and slime) provide plenty of excitement without the "dangerous" component. This is an important message especially for a chemical plant adjacent to neighborhoods. Everything is explained in terms the children can understand until the very end when we get to the Magic Cups experiment. By then they know it's not magic, but just some new science they will need (and want) more education to understand.

Safety Concerns

Safety and environmental concerns have become more important criteria than ever before. Older demonstrations are being revamped to use materials which can be disposed of down an ordinary drain or in the common trash. To whatever extent possible any materials which are corrosive or toxic are being eliminated and replaced with other compounds which will still enable the demonstrator to make the same points. In cases where there is no inert substitute the experiment materials are taken back to a chemistry lab for proper disposal.

Another safety-inspired trend is to use plastic or other substitutes for glass. This has several advantages. Obviously plastic is cheaper than glass, and when broken it usually doesn't shatter and is less likely to cause cuts. When carrying materials to the demonstration site the decreased weight of plastic becomes a motivating factor as well. Naturally the compatibility of the chemicals and the plastic must be tested, but usually inexpensive clear polystyrene drinking cups work very well for most demonstrations requiring beakers. Another substitution for glass is to use wooden sticks (popsicle sticks or bamboo skewers) or plastic spoons for stirring rods.

Finally, there are related considerations. The audience should be strongly warned not to taste or consume any of their science experiments. It is therefore my preference to be very cautious (especially when presenting to young children) with any demos that look like food, use food or pretend to be a consumable (eg. "Iced Tea or Grape Juice?"[1], where chemicals have been added to tea to cause the color change). On the other hand, touching and smelling may be quite instructive, so advice on when and how are quite useful. Teaching them at an early age to wash their hands after experimenting and not to put anything in their mouth while experimenting are habits which will always be practical. As a good role model you should follow these precautions, and of course wear your safety glasses while demonstrating, and provide them for students you invite to participate.

Getting and Utilizing Help

People to help with these projects can be found in many places. Within your own organization there are surely possibilities, but they may not be aware of their interest! Putting on the demonstrations for others at your place of employment will do a remarkable job of recruiting, especially if you are well organized and prepared to answer questions about how they can help. This method works equally well on both college students and professional scientists. Beyond the obvious sources you may want to look at professional organizations, such as the local ACS or AICHE sections for others who will work with you. People who have volunteered for other service work are good choices to ask, such as science fair judges or even Scouting leaders.

Prepare a handout to make it easy to use lots of help with various levels of expertise. In ours each experiment was given a number and a short key word title. A script of a few words with action steps noted was prepared. The handout also contains a table of reagents showing amounts per demo and how to prepare some of the solutions, a list of other supplies and comments, eg. suggestions on timing or safety, and references all keyed to the number and title of the experiment. Every volunteer gets the handout.

Another way to best utilize the volunteers is to have dedicated space and equipment for use in preparing for and doing the demos. We bought sturdy plastic boxes for "kits" and have a part of a room with the demonstration materials in it, with space for preparing solutions and clean up of the kits when they come back. Thus anyone involved in the project knows where to find what they need, and they can stop in if they have only a few minutes to help or pick up extra supplies quickly on their way out.

The volunteers may be divided by task. Some will not

want to make the actual demonstrations, or they may not have much time. These people can prepare reagents (by the gallon if there are a lot of demos going out) and organize the kits (pour up small quantities of reagents from the stock solutions and collect the other supplies), or turn the kits around after demonstrations (clean and repack). Also, since most of my volunteers are college students I find it helpful to use one of the more motivated or experienced students to oversee solution preparations and one to coordinate the other volunteers with contact persons at the schools who request demonstrations.

When all the demonstrations are ready and all the volunteers are on board it is time to go out to the public. It is necessary to make contact with teachers, principals, science coordinators or even the school district offices for permission to go into classrooms. It is not difficult to obtain the permission, but sometimes the lines of communication are weak as you wait for information to trickle down to the actual teachers. To get the school districts involved, Nalco invited the administrators and principals to their research facility and presented the demonstration to them. TAMUG contacted the school district who in turn sent a letter to each school and supplied the principals' names and phone numbers. We began in spring 1989, and Nalco visited 2500 students, while TAMUG saw about 50. For the '89-'90 school year TAMUG put on demonstrations for 700 and Nalco for almost 5000 in the Houston area. Nalco is now also demonstrating in the Chicago area, working out of another facility with a new set of coordinators. Word of mouth really keeps the program expanding.

Other Ways to Impact Students

As detailed previously, we made a conscious decision to visit elementary school classrooms. However, other ideas are worthwhile and should get some attention. For example, teacher-in-service programs are an excellent way to bring science into the classroom. While we do not have an in service program per se, at TAMUG there are similar programs for providing teachers with new information, such as the Sea Camp activities during the summer. This particular workshop teaches marine biology and field work, where both students and teachers can participate. The teachers can thereby earn one hour of graduate school credit.

An idea I discovered in a local elementary school has merit, i.e. to set up "science centers" in the classroom. Elementary students from kindergarten through fifth grade are encouraged to visit activity centers during their free time, or as a reward for good behavior. In the science center are simple experiments (each in separate baggies) with instructions so the kids can do the science for themselves at their own pace. The idea is to choose experiments which are safe, require essentially no supervision, and give the students enough to do to enjoy discovery, learn some principle and draw conclusions on their own. This type of center requires a volunteer to get it started and maintained, repacking when supplies are depleted or lost, and searching for new ideas to add.

Acknowledgments

Many thanks to the people around the country who have expressed an interest in our project, and especially to those who have shared their materials with me. My own children, Kristy and Lisa, have been our test subjects, both willing, eager and insistent. I consider this project to be a fun activity, and it has been supported, tolerated and even encouraged by my department. A special thank you to all the people at Nalco, especially Wes, who continue to expand their outreach and still have resources to share with us.

Literature Cited

1. Bailey, P.S., et. al. J. Chem. Ed., **1975**, 52, 524-525.
2. Madea, J. Science for Kids; Broward Community College - North Campus, 100 Coconut Creek Blvd., Pompano Beach, FL 33064.
3. Shakhashiri, B.Z., Chemical Demonstrations: A Handbook for Teachers of Chemistry; The University of Wisconsin Press, 1985, Vol. 2, p. 262.
4. Summerlin, L.R. and Ealy, J.L., Chemical Demonstrations: A Sourcebook for Teachers; American Chemical Society, 1985, Washington, D.C.
5. Ref. 3, p. 24.
6. Ford, L.A., Chemical Magic; T.S. Denison and Company, Minneapolis, MN, 1959. (Out of print).
7. McEvoy, J.E., Synergy, **1989**, 4(1), 1.
8. Summerlin, L.R., Borgford, C.L. and Ealy, J.B., Chemical Demonstrations: A Sourcebook for Teachers; American Chemical Society, 1987, Vol 2, Washington, D.C.
9. Ref. 3, Vol. 1, p. 31.
10. Cravey, M.J., submitted to J. Chem. Ed., **1990**.
11. Ref. 8, p. 95.
12. Rodriguez, F., Mathias, L.J., Kruschwitz, J. and Carraher, C.E., J. Chem. Ed., **1987**, 64, 72.
13. Ref. 9, p. 213.
14. Katz, D.A., Chemistry in the Toy Store; Department of Chemistry, Community College of Philadelphia 1700 Spring Garden Street, Philadelphia, PA 19130. (also available from the American Chemical Society).
15. Ligon, Jr., W.V., J. Chem. Ed., **1987**, 64, 1053.
16. Baker, P., Maryland Science Center, 601 Light Street, Baltimore, MD 21230, (301) 685-2370.
17. Barnett, B.E., Pollock, J.A., and Battino, R., J. Chem Ed., **1986**, 63, 460.

APPENDIX I
ADDITIONAL RESOURCE GROUPS AND MATERIALS
(in alphabetical order)

Alyea, H.N. and Dutton, F.B., Tested Demonstrations in General Chemistry, Journal of Chemical Education, American Chemical Society. Any materials by Alyea are of interest.

American Chemical Society: especially the Division of Chemical Education, the Committee on Chemical Education and the Joint Polymer Education Committee of the Divisions of Polymer Chemistry and Polymeric Materials: Science and Engineering 1155 Sixteenth St., NW Washington, DC 20036 (202) 872-4600. Excellent sources of materials and people.

Chem 13 News, Department of Chemistry, University of Waterloo, Waterloo, Ontario, Canada N2L 3G1.

Gardner, R., Kitchen Chemistry; Julian Messner, 1988.

GEMS (Great Explorations in Math and Science); a series available from Lawrence Hall of Science, University of California, Berkeley, CA 94720 (415) 642-7771.

Herbert, D., Mr. Wizard's Supermarket Science; Random House, 1980.

Herbert, D., Mr. Wizard's Experiments for Young Scientists; Doubleday & Co., 1959.

Herbert, D. and Ruchlis, H., Mr. Wizard's 400 Experiments in Science; Book-Lab, 1968 (Revised in 1983 by D. Goldberg), North Bergen, NJ.

Johnson, M., Chemistry Experiments; Usborne Publishing, 1981.

Journal of Chemical Education, Division of Chemical Education of the American Chemical Society, Subscription Department, 20th and Northampton Sts. Easton, PA 18042. See "Tested Demonstrations" and "Overhead Demonstrations" columns, in particular.

Katz, D.A., "Science Demonstrations, Experiments, and Resources", Department of Chemistry, Community College of Philadelphia 1700 Spring Garden Street, Philadelphia, PA 19130. An excellent list of resource materials.

National Science Resources Center, Arts and Industries Building, Room 1201, Smithsonian Institution, Washington, D.C. 20560. Focus is on teacher resources for elementary grades.

National Science Teachers Association, 1742 Connecticut Avenue, NW Washington, D.C. 20009. Several publications, with materials geared toward different age students.

Palder, E., Chemistry Magic; Woodbine House, 1987, Kensington, MD.

Schibeci, R.A., Education in Chemistry, **1988**, 25, 150-153. Another excellent collection of resource materials.

Smithsonian Family Learning Project, Science Activity Book; Galison Books, GMG Publishing, 1987.

Strongin, H., Science on a Shoestring; Addison-Wesley, 1976.

VanCleave, J.P., Chemistry for Every Kid; John Wiley & Sons, 1989.

VanCleave, J.P., Teaching the Fun of Physics; Prentice Hall Press, 1985. Includes science fair ideas.

Walpole, B., 175 Science Experiments to Amuse and Amaze Your Friends; Random House, 1988.

RECEIVED April 5, 1991

Chapter 13

Partners for Terrific Science

Creating Synergy Among Industry, Academia, and Classroom Teachers

Richard J. Sunberg[1], Arlyne M. Sarquis[2], John P. Williams[3], and Douglas B. Collins[2]

[1]Procter & Gamble, Miami Valley Laboratories, Cincinnati, OH 45239–8707
[2]Miami University Middletown, Middletown, OH 45042
[3]Miami University Hamilton, Hamilton, OH 45011

The *Partners for Terrific Science* program brings chemistry alive for classroom teachers and their pre-college students. The program is an industrial/academic partnership that brings teachers into contact with the science of their area's chemical industry. The program encourages and shows teachers how to use industrial-based chemistry to create hands-on science activities that students will find relevant, exciting, and interesting. By stimulating students' curiosity about the chemical world and by helping them appreciate their improved lifestyle gained through scientific and technological innovation, teachers and industry can promote "science excitement." Interested readers are encouraged to replicate the *Partners* program in whole or in part, depending on their needs and resources. For more information about *Partners,* please contact A. M. Sarquis, the program director.

Tom Runyan's 10th-grade students at Monroe High School (Middletown, Ohio) put 5 x 5 cm squares of uncoated, cold-rolled steel in culture plates, each lined with a vinegar-saturated piece of paper towel. They cover the plates and wait to see what changes occur. Within two hours or so, rust appears on the steel, signalling corrosion.

Runyan asks the students to think of ways to inhibit and possibly stop the rate of corrosion. His students suggest coating the steel with nail polish, lipstick, car wax, shoe polish, and tape. "Ideas," Runyan reflects, "that probably would have never entered my mind. I call it a 'brain explosion': when kids do hands-on science, they come up with fantastic ideas about how to apply and extend the experiments way beyond the original parameters."

The next day, the students apply some of their suggested coatings to the steel and repeat the experiment. After the students report their experimental

0097–6156/92/0478–0124$06.00/0

findings, Runyan discusses the chemical reactions that cause this corrosion and the methods and materials industry uses to inhibit it.

Helping Students Understand and Appreciate Science

Science teaching should help students appreciate and be more aware of the improved lifestyle gained from scientific and technological innovation. As Runyan admits, most of us take our quality of life for granted. We give little thought to the crucial role science and technology play in improving our lives. For example, we usually don't consider the improved integrity of our modern car's outer body until it eventually shows telltale signs of pitting and surface corrosion.

Science teaching should cultivate students' natural curiosity about the world around them. To foster this curiosity, students' initial contact with science should be an animated, hands-on experience, not an introduction to the dry equations that experienced practitioners use to manipulate science in the abstract. The *Partners for Terrific Science* program was designed with the belief that if pre-college students understand how chemistry and the chemical industry touch their daily lives, they will find chemistry more relevant, exciting, and interesting.

Runyan concurs with this belief, commenting, "When my students get their hands on an experiment and feel they have control over the experiment, I immediately have a more motivated and interested class than if I relied solely upon lectures or even demonstrations to teach them. I believe the number one problem for teachers is unmotivated students. I've made great strides in solving this problem by using the hands-on activities I've created with the *Partners* program. Through my work with the program, I've been able to give my students a special opportunity to participate in science as applied by locally important industries, industries that may someday employ them."

Building Ties with Industry

The *Partners* program, an industrial/academic workshop for teachers of grades 4-12, helps teachers develop ties with local companies that apply science and technology in their research and to their products. Once in the program, teachers learn about the chemistry used by the industries, including how the companies apply chemistry in their processes and analyses, and what chemical products these companies make. The teachers also consider the challenges chemical companies must meet in developing chemicals and products that are environmentally friendly and safe for consumers to use.

With guidance from industrial scientists from these companies, teachers implement innovative instruction and hands-on activities that will help students understand the impact of chemistry on their daily lives. In turn, the program increases the industrial scientists' understanding of educators through their interaction with the teachers in the workshop. The industrial scientists and teachers share their views of one another, cooperate, and mesh their efforts to form viable partnerships to improve science education. In doing so, they form an

Tom Runyan's students note the effects of corrosion upon the steel in their culture plate.

After completing the first part of Runyan's hands-on corrosion experiment, his students label their culture plate.

effective team working to educate society's future scientists and decision makers, our children.

These founding industrial companies comprise the core of the partnership (some of the science, products, and activities the companies explore, experiment with, and use in the *Partners* program follow in parentheses):

1. Armco, Inc. (corrosion, reactivity of metals, alloys, and refining and use of iron);
2. DuBois Chemicals (laundry products, germicides, and personal care products);
3. Henkel Corporation, Emery Group (use of tallow, preparation and testing of soaps, polymers, and lubricants);
4. Marion Merrell Dow, Inc. (stability of compounds, chromatography, toxicology, DNA, and genetic engineering);
5. Mead Imaging (photochemistry, polymers and emulsions, and separation techniques);
6. Procter & Gamble (chemistry of foods and food additives, including stabilizers, leavening agents, flavorings, and fats and oils); and,
7. Quantum Chemicals (natural and synthetic polymers, biodegradable and photodegradable plastics, and chromatography).

Along with these founding companies, Miami University, the National Science Foundation, the Ohio Board of Regents, the Cincinnati Section of the American Chemical Society, and the Ohio Chemical Council are major sponsors of the *Partners* program.

Roger Parker, a senior research chemist at Marion Merrell Dow who works with the *Partners* participants, says, "For me, one of the most satisfying aspects of the program is the teachers' enthusiasm. A lot of the teachers do not have a strong science background coming into the program, but they're willing to learn. They really want to take some of the science they've been exposed to in the program and bring it back into their classrooms for their students."

Parker adds, "As a member of the chemical industry, I've also been gratified to see changes in the teachers' attitudes toward applied science. Unfortunately, I think sometimes the public in general considers the term 'toxic chemicals' to be one word. While in the program, the teachers see that Marion Merrell Dow's chemical research into pharmaceuticals is saving lives, not destroying them."

Jeff Little, manager of manufacturing industries at DuBois Chemicals and a one-year veteran of *Partners,* notes, "By participating in the program, I'm helping ensure the long-term growth of DuBois. DuBois is really concerned about the shortage of students studying the sciences; the company is, for instance, having trouble finding qualified entry-level chemists to do the research needed to create marketable products. My hope is that some of the students the *Partners* program is influencing will walk through DuBois' doors seven or eight years from now and ask to work as chemists."

Exciting Teachers and Their Students

One goal of the *Partners* program is to increase the teachers' enthusiasm for chemistry so they will transfer their "chemistry excitement" to their students. Runyan, for example, got some of his ideas for relating the concept of the chemistry of corrosion and reactions with metal from his work with Armco, Inc. Armco manufactures steel at its Middletown, Ohio, plant, which also includes a newly added electrogalvanizing line and is home to its corporate research facility.

Runyan adds, "Every time I attend a *Partners* meeting, the other teachers and I have 'brain explosions' of our own. The teachers, the *Partners* staff, and the industry people have so many fantastic ideas about implementing and using all of these hands-on activities. I get a big morale boost when I know the community and industry support what I do in the classroom. I'm pleased that an industry such as Armco cares enough to use their own resources and the expertise of their people to help me implement the teaching of chemistry and hands-on science."

Jim Bowen, a sixth-grade teacher of life science at Greendale Middle School in Lawrenceburg, Indiana, uses a novel, effective way to transfer "science excitement" to his students. Bowen, a *Partners* workshop graduate, has created a rap song entitled "It's a Chemical World" through which he relates the chemistry of the gases of the air, the chemistry of plants and photosynthesis, and how chemists use and create polymers to improve our quality of life. Bowen comments, "Working with the people from DuBois, Marion Merrell Dow, and Quantum, I better appreciate the role chemistry plays in my life - how, for example, researchers at Marion Merrell Dow create and manipulate molecules to form new medicines."

Telling how his work with *Partners* has helped excite his students about science, Bowen adds, "The kids really love my rap, they sing it all of the time. By doing so, they learn that every time they put on a raincoat, every time they wear a hat, they are wearing and benefitting from materials that consist of polymers." Bowen says, "I have to be careful to include at least one hands-on science activity per week in my classes; otherwise, my students become antsy and depressed. If I act solely as a dispenser of science knowledge, as opposed to being a facilitator for their curiosity about science, I lose their interest, and they lose their motivation."

Running the *Partners* Program

Each academic year, 72 teachers participate in the *Partners for Terrific Science* workshop, which consists of three 3-day meetings in the fall and one 2-day meeting in the spring. Prior to the fall meetings, the program staff place the teachers into six groups of 12, based on the grade levels they teach. Typical groupings include: grades 4-5, 6, 7-8, 9-12 general science, and 9-12 chemistry, with teachers of gifted and talented students comprising the final section. Teachers need no special science background to participate in the workshop, as the workshop leaders review relevant chemistry principles during the meetings.

Also, the program provides the participants with an optional, one-week review of the fundamentals of chemistry in the summer prior to the start of the program.

Participants in the *Partners* workshop receive tuition-free instruction and a stipend for books and travel expenses. In addition, teachers receive graduate credit in chemistry from Miami University and several hundred dollars in materials and supplies to use to help forward the objectives of the workshop in their classrooms. The *Partners* program also provides funding to the teachers' school districts to defray the cost of substitute teachers while their teachers attend the workshop meetings.

At the start of each of the first three *Partners* meetings, the program directors pair each of the six groups of teachers with a mentor team representing one of the six focal companies. A mentor team consists of three people: an industrial scientist from one of the six focal companies (the industrial mentor), a college science faculty member (the academic mentor), and a classroom teacher who is a workshop graduate (the peer mentor). Each mentor team introduces their group of teachers to the focal company through discussion, hands-on laboratory sessions, and a visit to the focal company's industrial facilities. The on-site visit includes presentations by industrial research scientists and by managers, engineers, and technicians involved in manufacturing the company's products and operating the company's plant. Sessions on integrating industrial materials and hands-on teaching activities round out each meeting. Because the program pairs the groups of teachers with a different focal company at each of the first three meetings, teachers interact with and get ideas from three different focal companies during the course of the *Partners* workshop.

Not surprisingly, the teachers also make and enjoy many informal contacts among the mentors and their fellow teachers involved with the other teams. While carpooling to the *Partners* meetings, having lunch, and relaxing during breaks, teachers share ideas and classroom experiences with hands-on science.

Between the third and fourth workshop meetings, each mentor team assists 12 teachers in developing innovative final projects they can incorporate into their classroom curriculum and share with their fellow teachers. When they work on their final projects, the 12 teachers do not necessarily stay in their original groups based on the grade levels they teach. The program directors try to match each teacher with one of the three mentor teams he or she has visited and found particularly useful to his or her classroom setting.

Debbie Black, a *Partners* participant and sixth-grade teacher at Gerke Elementary in Franklin, Ohio, worked on a final project entitled, "What Can I Do with a Chemistry Degree?" She did the project in cooperation with the mentor team from Procter & Gamble (P & G), led by Dr. David Henry of P & G and Dr. Patricia Koochaki of Raymond Walters College in Cincinnati.

Using the project in her current curriculum, Black has her fifth- and sixth-grade students write and illustrate a book either on how to become a chemist at P & G or on how to start a career in chemistry in general. She introduces her students to the project by inviting several P & G employees to her class to discuss how they create flavorings for cakes and icings for their Duncan Hines baked goods and flavorings for their Folgers coffee. During the presentation, the

Teachers participating in the 1988–1989 *Partners* workshop posed for a group photo outside the Hughes Hall chemistry labs at Miami University, Oxford, Ohio.

P & G employees also discuss their career paths. Black reinforces the scientists' visits by conducting experiments on flavoring, coffee, and, as Black says, "Why Ivory soap floats."

"The kids' enthusiasm for science has really improved since I've incorporated hands-on science and industrial-based activities into my teaching," Black adds. "They love science; they come to class excited, asking, 'What are we doing today?' and, 'What are we going to do tomorrow?'" Black comments that, "Now, when they experiment with the flavorings of various foods, the kids realize that chemistry is not irrelevant, that it's something that's going on all the time. When their parents are cooking in the kitchen, the kids realize that they're doing chemistry in their homes."

Black remains involved in the *Partners* program, working as a peer mentor. "I asked to be considered for a peer mentor position with the P & G group, since I enjoyed participating in the program so much," Black says. "As a peer mentor, I link the chemists and teachers. The teachers do hands-on science in the program and say, 'Wow, this is great! Now, how can I apply it in my class?' With my experience in applying hands-on activities I've learned in the program to my class, I can suggest ways they can do the same," Black adds. "I also help the industrial and academic chemists relate to the teachers and make them aware of the specific needs of their teacher groups. For example, the P & G group provides glassware to the high school teachers, but they now provide the teachers of the lower grades with plasticware, out of concern for the younger students' safety."

Helping Teachers Develop Partnerships

Through the workshop, the teachers develop valuable support networks with other teachers, project staff, and industrial scientists. The mentor teams and program directors expect and help the teachers to develop industrial-based, hands-on teaching materials so they can share these activities with other teachers in their districts through in-service programs. To this end, the participants prepare activity reports along with their final projects. These activity reports enable their fellow teachers to integrate industrial-based activities into regular classroom instruction. Experience has shown that on the average, every teacher that participates in the program reaches at least 20 more teachers through in-service programs run by the teacher's school district.

Black comments, "As a result of my work with the *Partners* program, I was able to create a nine-week chemistry unit for my school district. Now, every sixth-grade teacher and the approximately 200 sixth graders within Franklin County Schools uses this unit."

"When my district first decided to emphasize hands-on science, I was somewhat concerned, having never taught it," Black notes. "However, attending the *Partners* workshop reassured me and gave me confidence, because I saw that a lot of innovative, dedicated teachers were already successfully teaching hands-on science. And, equally important, they were happy to share their creative ideas about teaching hands-on science with me."

Debbie Black's students do a "quantitative analysis", relying solely upon their sense of taste to discover characteristics about the unknown foods.

Bowen has enjoyed similar success in sharing the teaching activities he's created as a result of his work with *Partners*. He says, "Teachers within the program were really excited about the rap song. They saw the rap song as a way to relate science on the students' terms; so far, I've sent 35 to 40 tapes of the rap song to interested teachers. In return, I've received creative lesson plans from them and established a great network for the other teachers and me to share information." Bowen's rap has become rather popular: the American Chemical Society asked him to "rap" for them in Washington D.C., during their 200th National Meeting.

Gauging the Program's Success

Is the program a success? Quantitatively, pre- and post-workshop questionnaires show positive change in the teachers' attitudes toward the chemical industry, toward scientific innovation, and toward science education.

Excerpted from responses from the questionnaires, Tables I-IV graph the changes in the attitudes of teachers who participated in the *Partners* workshop during the 1988-89 academic year. For example, Table I shows that before taking the workshop, 22% of the teachers felt chemicals were either extremely or very dangerous to the average citizen. However, after the teachers took the workshop, only 10.2% of them retained these beliefs about the safety of chemicals. Likewise, Table IV shows that before taking the workshop, 16.4% of the participants felt chemicals should be used only in chemistry classes or industrial laboratories. After the workshop, only 5.1% of the teachers felt this way.

Black says, "If I can show kids that *chemistry* is not a big, bad word, that chemistry is not just using test tubes and writing formulas for no apparent reason, I go a long way toward firing their imagination about science. By bringing in scientists and chemists from companies such as Armco and Procter & Gamble, and by having them talk about their work and their careers, I help the kids understand that science is relevant and immediate. The kids understand that each day, humans - people like their Moms and Dads - apply science to everyday life."

Bowen adds, "I'm a very environmentally conscious person. By working with the chemists and researchers at DuBois, Marion Merrell Dow, and Quantum in the *Partners* program, they have become more human to me. I realize they are not impudently creating products that harm the environment; rather, they are caring people, with families, with moral obligations to their community. As a teacher of science and as a member of the community, I can relate better to these companies when I think of them in terms of their people and their people's ideas, as opposed to considering only the products these companies produce."

Inspiring Future Scientists through Their Teachers

In recent years, our students and teachers have been criticized for lacking the will to win in a technologically competitive world.

Table I. How Dangerous Do You Feel It Is for the Average Citizen to Handle Chemicals?

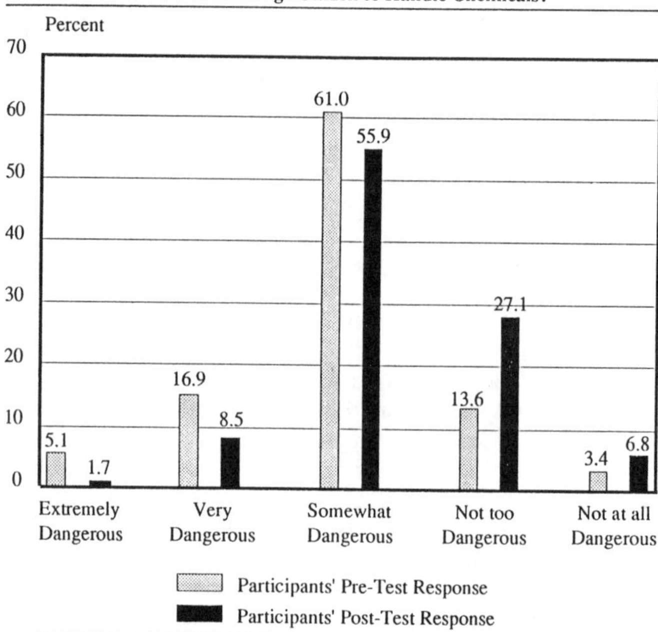

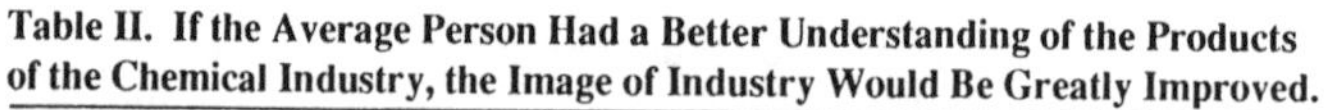
Table II. If the Average Person Had a Better Understanding of the Products of the Chemical Industry, the Image of Industry Would Be Greatly Improved.

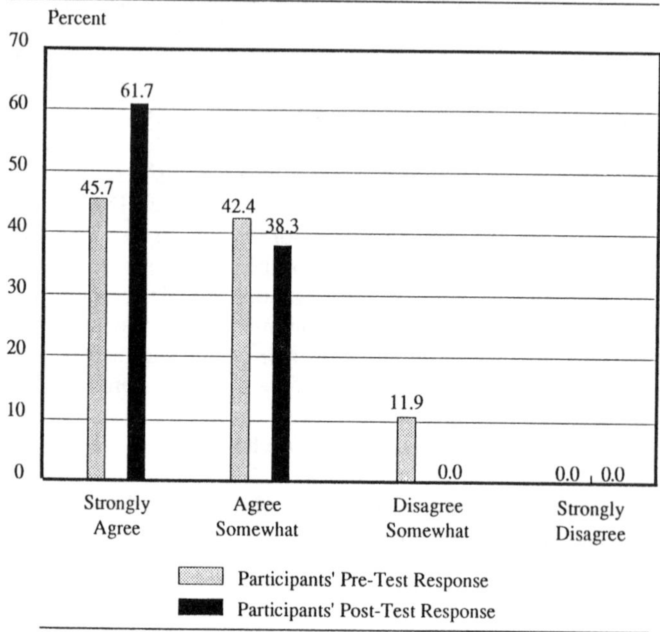

Table III. How Important is the Public's Understanding of the Chemical Industry for Them to Support It?

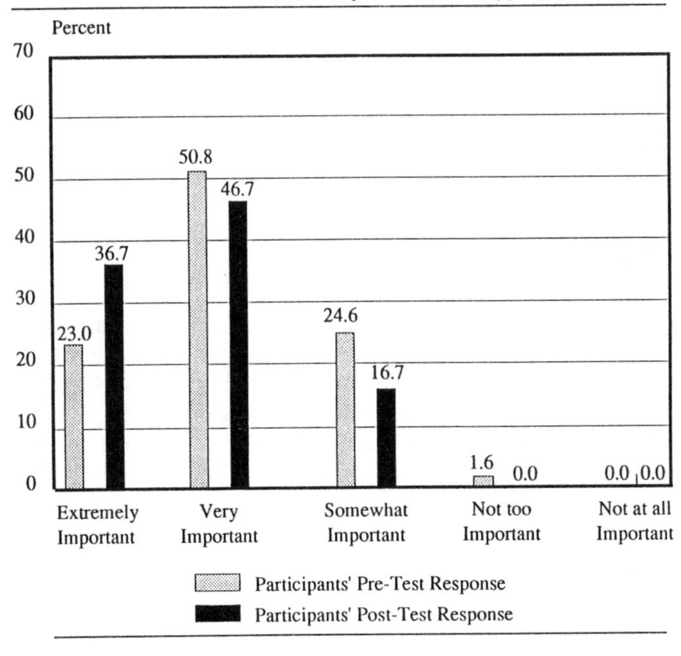

Table IV. Chemicals Should Only Be Used in Chemistry Classes or Industrial Laboratories.

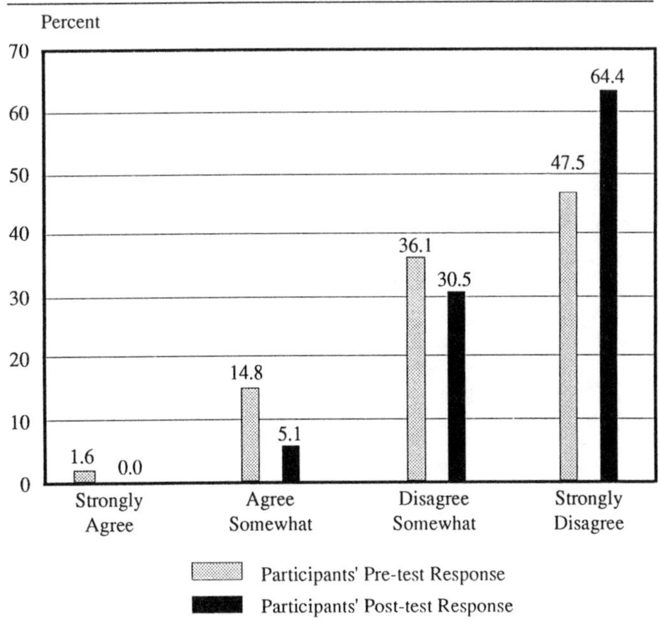

Rick Fayter, senior research director at Henkel Corporation, Emery Group, notes, "Observing the *Partners for Terrific Science* program, I was heartened by the teachers' spirit of innovation, enthusiasm, and dedication. These inspired teachers in turn inspire students; together, they have the will to meet the best challenge that global competition can offer." In satisfying their curiosity about our modern world, teachers and their students will better appreciate how applied science has contributed to its improvement. In appreciating the fruits of applied science, some of these students will be moved to make contributions of their own.

Starting a *Partners for Terrific Science* Program

The *Partners for Terrific Science* program works to bring chemistry alive for classroom teachers. Only when the teachers are excited about chemistry can the transfer "science excitement" to their students. *Partners* can work for groups of all sizes. All that is needed for a core program is one interested industry willing to work with one receptive school district. For more information about the *Partners for Terrific Science* program, contact Mickey Sarquis, the program director.

RECEIVED March 4, 1991

Chapter 14

Encouraging Tomorrow's Scientists Today

M. K. Snavely

BP Research and Environmental Science Center, Cleveland, OH 44128

BP's commitment to educational initiatives is international, as well as national and local, and expenditures total $15 to $16 million worldwide. This paper describes the international network which is being formed to coordinate the educational activities of BP's various business groups. The major programs of BP America, the U.S. arm of BP, are also explained, including a Scholarship-in-Escrow program for inner city schools in Cleveland, Ohio (headquarters of BP America); the development and testing of activities-based science materials for junior high use; Mathematics and Science collaboratives; and a partnership with one of the toughest Cleveland inner-city schools. These major programs, along with numerous options offered to individual students by the BP Research group, show the deep commitment and business involvement BP feels is necessary to encourage tomorrow's scientists today.

ENLIST, ENCOURAGE, EDUCATE -- these three words summarize the goals of the educational programs of BP, many of which are designed in particular to emphasize math and science at an annual cost of about $15 to $16 million worldwide. These programs occur in all parts of the world, since BP, the third largest oil company, has approximately 118,000 employees located in over 70 countries. However, since 40 percent of its assets are in the United States, many programs are focused here.

The company has always been noted for its commitment to education; in a recent "Visions and Values" statement, education has a significant role and is viewed as a priority for the company worldwide. This commitment can also be shown by the organization of BP's new corporate center in London, where educational affairs is a function coordinated by a small team of dedicated individuals responsible only to the COO and CEO and supporting the Managing Directors of the four business streams -- BP Exploration and Production, BP Chemicals, BP Oil, and BP Nutrition.

0097–6156/92/0478–0137$06.00/0

The main reason for BP's strong interest in education is cliched as "enlightened self-interest." BP believes educational activities provide strong business benefits on four fronts. They help to:

- Sustain BP's license to operate through improving BP's image and its relations with the local community,
- Meet short- and long-term recruitment needs at BP and improve the quality of people who apply for work,
- Extend the range of valuable information to which BP has access, particularly through its links with higher education,
- Encourage a thriving economy by improving education and enhancing understanding between the business and education communities.

Because BP is an international company, its educational interests are global, as well as national and local. A two-day conference was held in Brussels in November of 1989 to which all BP's main European subsidiaries sent a senior executive, who in turn brought a personnel or educational specialist from within their subsidiary plus two educators from their districts. The participants were told to devise an educational policy for BP that crossed national boundaries.

Almost too many ideas for future action flowed from the workshop sessions held during the symposium. BP Sweden, for example, proposed a summer college attended by school teachers from throughout Europe which would investigate solutions to common issues like motivating pupils, the introduction of new technology, and the decline in the number of young people. Deutsche BP suggested holding study weeks with pupils from one country visiting companies in another. Other ideas included international competitions for language learning among school children and a transnational clearing house for companies willing to give foreign pupils and students work experience.

These suggestions for programs, as well as all global involvement in education, are being examined and coordinated by BP's Educational Affairs (EDA) International Premier Network. The Network has recently announced that BP educational policies and programs should observe a number of fundamental principles expressed most simply by four words -- coherence, consistency, balance, and quality. The Network also announced that the programs would focus on two major themes:

- Science and Technology: Improving understanding of the natural and man-made world, the fruits of science and technology, and their use in a valuable and responsible way, and
- Environmental Education: Fostering understanding of environmental issues.

Included with this strong and growing international trend in education support is a segment dedicated to U.S. educational interests. Most of the U.S. educational activities are coordinated through the U.S. arm of BP -- namely BP America, formed in 1986 from the merger of BP interests in the U.S. and The Standard Oil Company (previously known as Sohio).

What is BP America doing in the U.S. to ENLIST, ENCOURAGE AND EDUCATE? As early as 1982, which was still The Standard Oil era, the Board of

Directors authorized programs for the improvement of public elementary/secondary education, particularly in those communities where the company had a significant presence. Since then, the U.S. corporation has made grants of approximately $1 million annually to improve education in local communities.

Because the headquarters of BP America is located in Cleveland, many of its contribution dollars focus primarily on Cleveland schools, which manifest all the problems typical of urban centers and whose potential human resource is usually largely wasted. The Cleveland schools are used as a laboratory to determine which educational experiments work and which do not work. Some funds are used to support national organizations, but even they are encouraged to use the Cleveland schools for demonstration purposes.

Some facts about the Cleveland schools: the population in the public school system is 75% minority, 70% of whom are black. More than 50% of its students are from single-parent homes; almost 70% of the children qualify for free lunches. The dropout rate is approximately 50%; attendance on any given day is only about 70%. The district is also under one of the strictest desegregation orders in the country, and a great deal of time and energy of the populace has been spent on avoiding compliance. There has been ongoing public disagreement between the superintendent and the school board, which has resulted in the last three superintendents having their contracts bought out.

In late 1987, BP America committed $1.9 million over five years to a new comprehensive, community-wide effort -- the Cleveland Initiative for Education (CIE). CIE was created by the Greater Cleveland Roundtable, a neutral forum where civic leaders from many walks of life come together to discuss critical community issues. CIE's goals are to improve the quality of Cleveland's high school degree, increase the number of graduates who go on to post-secondary institutions, and provide other graduates with the qualifications to fill entry level jobs in local business. Total financial goal of the program was $16 million for the Initiative's first five years -- $10 million designated for a Scholarship-In-Escrow Program and $6 million for a School-To-Work Program. By August 31, 1990, the campaign had raised 96% of its goal, with $15,407,569 in private funds pledged. Of that total, $7,874,319 was committed by the business community, $7,071,900 by foundations, and $461,350 by individuals.

It is important to note that BP's involvement has been much more than financial. In most cases, senior management and BP Corporate Contributions staff have been among the leadership in putting programs together and in shaping strategies for educational reform. Senior management has taken leadership positions in a number of the organizations involved, including the CIE.

Scholarship-in-Escrow Program

One of the major components of the CIE program is Scholarship-in-Escrow (SIE), in which students earn dollars for grades in core academic subjects.

Every student in grades 7 through 12 can earn $40 for every A, $20 for every B, and $10 for a C in core subjects (English, math, social studies, science, and foreign language). Students enrolled in Major Work/Honors classes earn a $10 bonus per grade. The money is held in escrow for use in post-secondary education when the student graduates from the Cleveland Public Schools. The maximum that can be earned is $6,000, which will not pay completely for a college education, but can motivate students to achieve and earn additional scholarships.

Students began earning SIE funds in February, 1988. By the end of the 1989-90 school year, $10,417,900 in entitlement to scholarships had been earned by 35,861 students. Students and their parents receive a cumulative quarterly statement of earnings, but the money is paid directly to the post-secondary institution selected by the student. SIE funds can be used only at Pell Grant- or Ohio Instructional Grant-approved two- and four-year colleges, technical, and vocational schools.

Thus far, SIE is yielding encouraging results; academic performance is improving. After the first three years, average SIE funds earned by 7th through 12th graders for good grades in core academic subjects are up 18.9%, even though enrollment is down 3.2%. The number of A's, B's, and C's earning SIE funds is up 13.2%.

But financial incentives alone are not enough to change the future for Cleveland Public School students; a student advocate program was also implemented. There is now an advocate in every comprehensive high school and intermediate school in the district. The role of the SIE advocate is to match students' needs to community resources. They work with individual students and groups to improve grades, help with college entrance processes, and plan cultural and educational experiences.

More encouraging statistics: as of August 31, 1990, $188,000 in earned SIE funds has been paid out for 1,159 Cleveland Public School graduates to attend 194 different colleges and post-secondary schools. While total enrollment in the 12th grade has declined since the class of 1988, the percentage of 12th graders graduating and using SIE funds is increasing. Five months after graduation, 420 members of the class of 1990 requested $50,800 of their SIE funds, compared to 134 members of the class of 1989 requesting $21,600 of their earnings during the same period last year.

Critical to assembling the full financial package needed to pay tuition, 24 colleges and post-secondary schools in Ohio and neighboring states have volunteered to match or exceed SIE funds earned by Cleveland Public School students attending their schools.

BP America/West Technical High School Partnership

As a complement to the SIE program, each corporation in the city was asked by the Superintendent to form a partnership with one of the comprehensive high schools. BP America agreed to partner West Technical High School, one of the largest and toughest schools in the district. This partnership is extremely people-intensive and its purpose is to revitalize the mission of the school and to motivate both staff and students. The various committees which are made up of BP representatives and West Tech teachers are shown in Table I below.

Table I. West Tech/BP America Education Partnership

Committees:

- Mentorship/Tutor Program
- Curriculum Enrichment
- Independent Study
- Enrichment Fund
- International High School
- Job Development
- On Task Liaison

Although there are many components to this partnership, one of the major ones is the mentoring program in which 50 BP employees are acting as mentors for 53 West Tech students. The BP mentors provide support and guidance to help their proteges persist in their studies, overcome stumbling blocks, and set realistic goals for their future. Requirements include weekly communication between mentor and protege and attendance at a minimum of seven scheduled special events and activities. BP America provides training, ongoing support, and supervision to the program participants. This program also supports a BP goal to encourage employees to become involved in education.

Another unique component of the BP America/West Tech partnership is the development of an International Studies Program (IS), which is particularly appropriate because the students represent a variety of ethnic groups: African American, European, American Indian, Asian American, and Hispanic. The International Studies Program consists of a four-year high school plan requiring students to complete a foreign language program and a social studies component centered on multicultural comparisons that stress interdependency among nations. In a project underwritten by BP America, twenty foreign professionals work with the IS classes throughout the school year in the areas of multicultural studies, world history, conflict resolution, comparative political systems, and languages.

Science Materials for Junior High Grades

The company's first venture into public education in mid-1982 was a $1.7 million commitment to the American Association for the Advancement of Science (AAAS) to develop and test activities-based science materials for junior high use. The intermediate school level was the targeted level because it is the critical gateway into further science and mathematics and yet the curricula seem the most intellectually barren. The hands-on science activities were designed to encourage development of critical thinking skills, to engage students actively in the learning process, and lure teachers away from exclusive reliance on textbooks. Incidentally, all BP America grants for curriculum reform include a major component in teacher training because, unfortunately, too many teachers at the elementary and junior high level are ill equipped educationally to teach math and science.

The materials produced by AAAS, known as Science Resources for Schools, were packaged in coordinated kits and sent to teachers several times a year. The experiments relied on easy-to-come-by and very inexpensive materials.

In addition to the AAAS project, BP America, since 1984, has also underwritten the development of science curricula units illustrating a number of critical scientific principles, currently being edited and published by the National Science Teachers Association (NSTA).

One of the workbooks, Earth: The Water Planet,*(1)* deals with a number of important concepts in earth science that are taught in grades 7 through 9. Another, Methods of Motion*(2)* (an introduction to mechanics), deals with physical science concepts and is intended for the same grade levels. BP America has underwritten two more workbooks in this series: a second book on mechanics, Evidence of Energy,*(3)* and one on electricity, which will be published soon. An additional BP grant has been made to the NSTA to develop four activities-based workbooks in a series on earth science topics.

The development of these materials tries to address two critical needs cited by the experts:

1. Exemplary hands-on science materials for the middle grades in order to interest students in further study when they move on to elective subjects;
2. In-service training for middle school science teachers, who generally have poor backgrounds in science. These materials convey important concepts in a way that teachers can understand and integrate into classrooms.

The National Science Teachers Association (NSTA), which is the largest service organization for the country's science teachers, runs workshops on the materials at their regional meetings and their national meetings. NSTA also promotes the sale of the workbooks and features them in their publications catalogue. We are told that BP America's books are best sellers among the more than 200 titles they publish, and demand for the workshops is so strong that teachers have to be turned away. We also know that the workbooks have been selected as exemplary by several national organizations working to improve science education.

The company has also sought out and underwritten curriculum development/teacher training projects at The Ohio State University for 7th and 8th grade pre-algebra units and for 12th grade pre-calculus courses. In all of these cases, Cleveland teachers are given in-service training and are part of the pilot testing of the materials.

Participation in Science/Math Collaboratives

As BP America gained experience in working to achieve educational reform, we learned very quickly about the advantages of partnerships and the value of intermediary organizations. We could sit around the table and work on a common agenda with representatives of other businesses, higher education, and the school district, and we learned that a great deal more could be accomplished than if we attempted to "go it alone."

One of these intermediaries is the Cleveland Education Fund, created in 1984 largely with Ford Foundation and Cleveland Foundation money. The mission of the fund was to motivate teachers and pilot promising programs for the Cleveland School District. It resulted in Cleveland successfully competing and obtaining one of the five major grants that the Ford Foundation made in 1985 for the creation of a Mathematics Collaborative. The success of that collaborative attracted a Carnegie Corporation grant to develop a Science Collaborative with parallel goals. Senior managers from BP participate on the Collaborative advisory committees and on the Board of Trustees of the Cleveland Education Fund.

Both the Mathematics and Science collaboratives have been very successful, and many of the activities are listed in Table II below.

Table II. Components of Math and Science Collaboratives

- Summer Internships
- Scholarships for Graduate Study
- After-school Workshops
- Symposia Sponsored by Industry
- Professional organization Meetings
- Small Grants for Innovative Classroom Methods
- Participation in National Science/Math Education Networks
- Resource Center
- Newsletter

BP Research Educational Initiatives

Educational Initiatives coordinated internationally by the BP Educational Affairs and in BP America have been described, but no mention has been made about specific activities of our BP Research Center, recently renamed the Warrensville Research and Environmental Science Center. The center has approximately 700 employees (450 scientists) and is located in Warrensville Heights, a southeastern suburb of Cleveland. There are quite a variety of active educational programs as shown Table III, and it will only be possible to highlight a few.

Initiatives include a minority intern program, a teacher intern program, shadowing programs for students to follow researchers in the laboratory for a day, and participation in the mentoring program with West Technical High School.

Also, career day presentations and school group tours are given, participation in the career activities of the Women In Science and Industry Organization is encouraged, and a unique Explorer Scout program is underway. The latter is a fledgling co-ed organization associated with the Boy Scouts of America which one of our Senior Scientists originated because of his dedication and interest in these young people. Approximately 20 high school students from the area meet twice a month at the Warrensville Research facility, where programs are designed to pique the curiosity of these young people and to encourage their interest in science.

BP Research also sponsors high school senior projects which usually consist of two- to three-week projects completed under the direction of one of BP's scientists. Employees also act as Science Fair judges and Mathcounts volunteers (a national competition for junior high age students).

BP Research gives science fair awards each year to the best projects in regional science fairs. The winners receive a plaque, a monetary reward, a luncheon and tour of the laboratory, and an opportunity to display the project at the laboratory for several weeks.

In 1989, a new initiative was tried. The Northeastern Ohio Educators' Association (NEOEA) always holds an in-service day in October. BP Research decided to sponsor a symposium for junior and senior high school science teachers called "Science and Technology At Work In Industry." One of the main purposes was to illustrate to the teachers how the scientific principles they taught in the schools were put to use in industry. Topics included energy, analytical problem solving, environmental concerns, and artificial intelligence. The symposium was very well-received and it was repeated in 1990.

In addition to current programs, there are exciting possibilities for the future. BP's Educational Affairs Network, mentioned previously, is becoming more active, and members are beginning to communicate and take advantage of each other's resources. The resulting synergistic effect has almost limitless possibilities.

ENLIST, ENCOURAGE, EDUCATE -- it is heartening that our own internal business groups along with external businesses are realizing they must participate

Table III. BP Research Educational Initiatives

• Minority Intern Programs	• Career Day Presentations
• Teacher Intern Programs	• School Group Tours
• West Tech High School Mentoring Program	• Explorer Scout Program
• Shadowing Programs	• Women in Science and Industry (WISEMCO)
• High School Senior Projects	• Science Fair Awards
• Science Fair Judges	• NEOEA Day Symposium
• Mathcounts Competition Volunteers	• Special Programs

and get involved with educational initiatives -- there is no choice. Company managers are recognizing that educational initiatives definitely have a place in company strategies and are one of our strongest preparations for future business competition. We're all learning together what works and what doesn't work, and, hopefully, we can expand on our successes and stimulate others to an awareness that TOMORROW'S SCIENTISTS MUST BE ENCOURAGED TODAY.

Acknowledgments

This paper included information from three sources:

1. Thomas, David, "Oiling the Wheels of Education," Financial Times, December 1, 1989, p. 13.
2. Hardis, Sondra, Manager, Education Programs, BP America Inc., 200 Public Square, Cleveland, OH, 44114-2375, personal communications.
3. "Scholarship-in-Escrow, The Third Year," Scholarship-in-Escrow, 1380 East Sixth Street, Room 312, Cleveland, OH 44114.

Literature Cited

(1) Earth: The Water Planet, National Science Teachers Association, 1742 Connecticut Avenue N.W., Washington, DC 20009, Stock Number PB75.
(2) Methods of Motion, ibid, Stock Number PB39.
(3) Evidence of Energy, ibid, Stock Number PB80.

RECEIVED April 29, 1991

Chapter 15

Reactivity Network

Cooperation Between Academic Inorganic Chemists and High School Teachers

E. K. Mellon and T. G. Berger B.

Department of Chemistry, The Florida State University, Tallahassee, FL 32306–3006

The aims of the Reactivity Network are to collect descriptions of chemical phenomena and suggest how these phenomena might be transformed into experimental problems for K-12 students. Network writing teams are made up of content experts and experienced teachers whose goal is to produce reviews to be published in the Journal of Chemical Education. The reviews will be reprinted and disseminated to chemistry teachers nationwide through a network of about 250 chemical educators at all levels from elementary through college. Use of local networks expands the reach of the Reactivity Network to thousands of teachers. These reviews stress descriptive chemistry, for example, reaction rates, driving forces, and catalysis in the context of student's everyday experiences. The Network examines alternate methods of presenting material which might pose waste disposal or health problems if performed in a traditional manner. Interesting examples of information gathered by the Network are presented.

"...Scientists seldom invent a theory without having first an anomalous or puzzling event to explain. It seems odd to me that we should expect students to learn science without being puzzled first, or without even having asked a scientific question..." E. M. Vitz(*1*)

The idea for the REACTIVITY NETWORK(*2*) began as early as 1978 at an international conference on introductory chemistry entitled "New Directions in the Chemistry Curriculum"(*1*) held at McMaster University. Participants in this Conference agreed that the general chemistry courses at both the high school and college levels are overloaded with theory. Worse still, this oversimplified theory is presented to an audience insufficiently mature to

0097–6156/92/0478–0146$06.00/0

appreciate it. Participants proposed, rather, that beginning chemistry courses should contain an artful blend of appropriate theory and the direct observation and interpretation of chemical phenomena. It would be best if these observed phenomena were concrete, and rooted in the student's everyday lives. For example, inorganic chemical reactivity (What reacts with what? How far? How fast?) is far under-represented in our beginning courses where the emphasis is on "...the uncomprehending memorization of over-simplified theories."(*1*) The REACTIVITY NETWORK was organized as one mechanism towards the reintroduction of chemical reactivity into beginning chemistry courses.

The REACTIVITY NETWORK, directed by E. K. Mellon, began formally in early 1987 with a planning conference supported by the ACS Society Committee on Education. Support for this project by the NSF began in the summer of 1987. The NETWORK is dedicated to reducing the endless mass of inorganic chemical reactivity information in the chemical literature into a form usable by teachers, curriculum developers, and textbook authors. Inorganic chemical reactivity was chosen as the primary focus of the REACTIVITY NETWORK because it provides colorful, interesting phenomena with which to rivet student interest, and because it yields a rich bounty of experimental problems at all levels for students to solve.

Network writing teams are comprised of content experts and experienced teachers whose goal is to produce reactivity reviews for publication in the Journal of Chemical Education. Simple reactions, driving forces, rates, catalysis and the manipulation of equilibrium states by temperature and concentration changes are stressed---all within the context of relating the reactions to student's everyday experiences. The reviews will be disseminated to chemistry teachers nationwide through a network of some 250 chemical educators recruited at all levels from elementary school through college. The high school teachers recruited for the Network have their own local teacher networks, so that the REACTIVITY NETWORK has the potential of reaching thousands of high school chemistry teachers.

The first of the reviews is a survey of the reactivity literature dating to Michael Faraday in the 1820's.(*3*) The second by a team of Arizona teachers led by Jim Birk(*4*) encompasses the reactivity of nickel, and is typical of the reviews to come, spanning in student interest the middle school years through first year college chemistry. The third review, authored by a Georgia team led by Butch Atwood, addresses the difference between inorganic and nuclear reactivity, and the reason why entropy change is important in one domain and not the other.(*5*) This paper is intended for advanced placement or college students.

The reviews will encourage the use of microscale manipulations since chemical safety and waste disposal are of such grave public concern. They will call for the use of equipment, such as chemical microscopes, already present in most schools and colleges but under-utilized in chemistry instruction. At the urging of the American Chemical Society Committee on Education, sensitive areas, such as chromium and nickel chemistry---where many school systems have banned these reagents altogether---will not be sidestepped. The more hazardous operations, which are clearly identified as such in the reviews, may well find their way into the high school laboratory by way of modern teaching technology, for example on videodisc or videotape.

The recent American Association for the Advancement of Science report Science For All Americans(6) calls (once again) for extensive reforms in our educational system, suggesting that when we teach for scientific literacy, the instruction, to be honest, must be consistent with scientific values and with the spirit and character of scientific research. Why stress the memorization of answers? Should we not stress the framing and answering of scientific questions, instead? Is not this the procedure we use in our research? Should not much of students' instructional time be spent on evidence collection, hypothesis generation and problem solving? Should we not reward curiosity and creativity instead of unquestioning memorization?

Specifically, Science For All Americans(6) reminds us that what the lecturer projects is not necessarily what the student learns. Meaning must be constructed in the mind by each student individually. Frequently, students must radically restructure thinking to banish misconceptions. This process requires working with concepts over time in a variety of situations, preferably problem solving ones. The REACTIVITY NETWORK reviews will collect descriptions of chemical phenomena and suggest how the phenomena might be transformed into experimental problems.

Learning progresses from the concrete to the abstract. Perhaps our greatest fallacy is to mistake students' familiarity with jargon as true learning. Is is time to pare down the number of topics covered and emphasize only the more important. Quality learning is time consuming.

A second major curriculum reform project coming on the scene is the National Science Teachers Association's "Scope, Sequence, and Coordination of Secondary Science" (SSC) project. It seeks to attack the problem of science illiteracy by having every student study science (no tracking) in grades 7-12. Chemistry instruction will be spread out over several years time, and will progress from the concrete to the abstract.

The ACS white paper "Education Policies for National Survival(7) suggests that for grades K-8 we need "...the development of safe, hands-on laboratory experiences that present science as a problem-solving endeavor within a societal context..."

Everyday common sense leads one to the conclusion that laboratory instruction should be dedicated in large part to the solution of experimental problems. Before true problems can be addressed, however, the interest of the student (and often, of the teacher) must be captured. The chemistry sets which made chemistry so attractive to youngsters in years past are now history. The REACTIVITY NETWORK Project will reintroduce many of those engaging chemical phenomena.

Here are some interest capturers collected from the REACTIVITY NETWORK. The supplies are readily accessible and the procedures simple. Most chemists will be unable to resist trying them:

** Reversibility is demonstrated in this first example. Nickel(II) nitrate hexahydrate is placed in a cryophorus--a sealed tube(4,8) and thermally decomposed. If the container is allowed to sit undisturbed, the $Ni(NO_3)_2{\bullet}6H_2O$ is regenerated.

Good demonstrations raise experimental questions. Can other metal nitrates be used? What about other hydrates? Are there cases where regeneration does not take place? What has happened to the metal ions in the cases where regeneration does not take place?

PROCEDURE: Make a U-tube out of 8mm glass tubing and seal one end. Chase the moisture out of the tube with a flame or heat gun. Allow the tube to cool and add a few crystals of blue $Ni(NO_3)_2{\bullet}6H_2O$. Attach the open end to a water aspirator and evacuate. Seal the tube with a flame. Taking appropriate safety precautions, gently heat the nickel(II) nitrate crystals. The thermal decomposition produces black nickel(II) oxide in the tube, water vapor on the walls, and red gaseous NO_2. Blue crystals of the hydrated nickel(II) nitrate begin to reappear after a few hours. The nickel complex is observed to be fully regenerated in several days.(*4*) With care, this system can be reused many times, thus no waste is generated.

** In this next relatively simple example mineral-like crystaline layers grow in a test tube. Water is added to a large test tube charged with iron nails and layers of salt and copper(II) sulfate. The nails are oxidized, while the copper ions are reduced to the metal. The products form in aerobic and anaerobic regions of the tube. A beautiful copper tree grows from the iron nails downwards into the salt layer. Although the first changes results are observable within a few minutes, the process develops over a period of months. It is rare that students encounter any system that takes more than an hour to reach equilibrium. It is best to leave the system undisturbed. Students can assemble and store the tubes at home because of the innocuous nature of the components.

Will other metal-salt combinations work as well? Will such systems be speedier? Slower? What is the composition of each layer? Why does not the copper metal tarnish even after long periods of time? What is the role of the salt layer? What is the nature of the airtight plug? In the anaerobic portion of the system what is being organized into "mineral strata"? Do all iron-copper(II) sulfate tubes react at the same rate? Can the strata be separated for qualitative analysis?

PROCEDURE: Add 1" of solid $CuSO_4{\bullet}5H_2O$ to a large, dry test tube. Add a circular filter paper spacer. Now add 2" of iodide-free NaCl and another spacer on top. Place two or three ungalvanized iron nails on top of the second spacer. Carefully, add water to cover the nails (ensure there are no entrapped bubbles). Leave undisturbed and observe the changes. No special precautions with respect to waste disposal need to be taken with this experiment/demonstration.

SUGGESTION: A rack may be built to hold a number of test tubes. Each year the students set up a fresh tube and compare it to the results observed in the previous years' experiments. As indicated previously, initial results are observable within a few minutes.(*9*)

** This demonstration involves painting with compounds that change color when they are dried with a heat gun. It is a variation of the pink-blue "disappearing ink" trick using hydrated/dehydrated cobalt(II) chloride.

QUESTIONS: Why is there a change in color when the salts are dried? If they are allowed to reabsorb water, do they return to their initial colors? If shades of coloration are desired, which pigments should be applied first? Why will the paintings containing iodides yellow if allowed to remain hydrated? Does this also happen with the iron(III) chloride? What would happen if the paper is kept dry, in bottles, with a desiccant present? Can other metal salts be used for different colors? Will they also change color in time? Will different metals interact to yield results you would not expect?

PROCEDURE: Dissolve mixtures of cobalt(II) chloride and various quantities of sodium or potassium bromide and iodide in water. Iron(III) chloride can also be used. Use a paint brush to make water color paintings on paper (filter paper works very well). With some experimentation, water color paintings which encompass a wide variety of colors (yellow, pink, green, blue) can be made which appear when heated and disappear in a humid atmosphere.(*10*) These solutions can be kept for the next year, or, since they contain very little cobalt, disposal in the trash is appropriate. The salts can be washed off the paper and the cobalt recovered from the resulting solution.

Most importantly, reactivity phenomena can form the bases for experimental problems. The fact that CuCl is a white, insoluble solid is certainly memorizable. How much more meaningful is this fact when it is the key to the solution of a problem: Place 1 cm pieces of bright copper wire in each of two small test tubes. To one tube add 0.5 mL of dilute $CuCl_2$ and to the other add 0.5 mL of dilute $CuSO_4$. The problem is to account for the difference --in the tube where chloride is present, the copper corrodes away, and a white solid (CuCl) is formed. In the other tube the Cu remains bright. The formation of insoluble CuCl drives the corrosion reaction forward.

Textbooks tell students that "ferrous hydroxide is a pale green solid." As Eugene Rochow pointed out during the 1978 McMaster Conference(*6*), the dry recital of memorized fact is what killed the "old" descriptive chemistry. Anyway, pure ferrous hydroxide is not pale green, it is white. But when one mixes ferrous and hydroxide ion solutions in the open air, the precipitated green ferrous hydroxide darkens rapidly to a black color, and eventually becomes the familiar red-orange hydrated ferric oxide. This phenomenon raises many questions, all answerable experimentally:

*What causes the decolorization?

*How would you make white $Fe(OH)_2$?

*Can you prevent (mask) ferric hydroxide precipitation by adding ligands known to complex the ferric ion?

*Does the rate of air oxidation of ferrous ion vary with hydroxide ion concentration?

This iron chemistry phenomenon can be related to day-to-day concerns:

Why is window glass green when viewed edge on? What evidence do we have from iron chemistry concerning the oxygen content of the earth's prehistoric atmosphere?

We can no longer consider the accurate transmission of content by lecture alone to constitute all that is required for effective chemistry teaching at the high school and grade school levels. The student market has moved in a different direction and we must alter our methods accordingly.

Acknowledgment. The REACTIVITY NETWORK is sponsored by the U.S. National Science Foundation under Grant Number: MR 87-51183.

Literature Cited.

(1) Proceedings of an International Conference on Introductory Chemistry: New Directions in the Chemistry Curriculum; McMaster University: Hamilton, Ontario, Canada, 1978.
(2) Mellon, E.K. SYNERGY, 1990, 4, No. 2, 6-7.
(3) Mellon, E.K. J. Chem. Educ., 1989, 66, 251-256.
(4) Birk, J.P.; Bennett, I., Boran, M., Kinney, C. J. Chem. Educ., 1991, 68, 48-54.
(5) Godfrey, I; McLauchlan, R.; Atwood, C.H. J. Chem. Educ., accepted for publication.
(6) Science for All Americans: A Project 2061 Report; AAAS: Washington, D.C., 1989.
(7) ACS White Paper: Education Policies for National Survival, November 14, 1989, American Chemical Society, 1155 Sixteenth Street, N.W., Washington, D.C. 20036.
(8) Campbell, J.A. J. Chem. Educ., 1970, 47, 273-277.
(9) Cortez, J.A.; Powell, D.; Mellon, E.K. J. Chem. Educ., 1988, 65, 350-351.
(10) Bare, W.D.; Mellon, E.K. J. Chem. Educ., submitted for publication.

RECEIVED April 1, 1991

Author Index

Affiliation Index

Subject Index

B

C

D

E

O

P

Q

R

S

Production: Peggy D. Smith
Indexing: Colleen Stamm
Acquisition: Cheryl Shanks
Cover design: Alan I. Kahan

Printed and bound by Maple Press, York, PA

BC